Systems Engineering - Design, Analysis, Programming, and Maintenance of Complex Systems

Edited by Germano Lambert-Torres, Gilberto Capistrano Cunha de Andrade and Cláudio Inácio de Almeida Costa

Published in London, United Kingdom

Systems Engineering - Design, Analysis, Programming, and Maintenance of Complex Systems
http://dx.doi.org/10.5772/intechopen.110997
Edited by Germano Lambert-Torres, Gilberto Capistrano Cunha de Andrade and Cláudio Inácio de Almeida Costa

Contributors
Abdul Nazeer, Adeel H. H Suhail, Alebachew Mengistu Worku, Amin Azad, Arjuna Marzuki, Boris Kontsevoi, Farnaz Khodaei, Fesiha M. Guangul, Filippo Ghiglieno, Hananneh Ardestani, Ikechi A. Ukaegbu, Mequanent Ewnetu Belay, Olavo Leopoldino da Silva Filho, Paulo Henrique Dias Ferreira, Peter Schlosser, Robert F. Bordley, Roya Gazori, Tamrat Yifter Meles, Thomas Kusel, Tibor Schlosser, Vinicius Tribuzi, Yewondwosen Gzate Ayalew

First published in London, United Kingdom, 2024 by IntechOpen
IntechOpen is the global imprint of INTECHOPEN LIMITED, registered in England and Wales, registration number: 11086078, 167-169 Great Portland Street, London, W1W 5PF, United Kingdom

British Library Cataloguing-in-Publication Data
A catalogue record for this book is available from the British Library

Additional hard and PDF copies can be obtained from orders@intechopen.com

Systems Engineering - Design, Analysis, Programming, and Maintenance of Complex Systems
Edited by Germano Lambert-Torres, Gilberto Capistrano Cunha de Andrade and Cláudio Inácio de Almeida Costa
p. cm.

This title is part of the Industrial Engineering and Management Book Series, Volume 7
Topic: Production Engineering
Series Editor: Fausto Pedro Garcia Marquez
Topic Editor: Orhan Korhan

Print ISBN 978-0-85014-339-3
Online ISBN 978-0-85014-340-9
eBook (PDF) ISBN 978-0-85014-341-6
ISSN 3029-0511

For EU product safety concerns:
IN TECH d.o.o., Prolaz Marije Krucifikse Kozulić 3, 51000 Rijeka, Croatia,
info@intechopen.com or visit our website at intechopen.com.

IntechOpen Book Series

Industrial Engineering and Management

Volume 7

Aims and Scope of the Series

Industrial Engineering and Management (IEM) is a discipline that focuses on optimizing complex processes and systems within various industries. It involves the integration of engineering, business, economics, mathematics, and behavioral sciences to improve efficiency, productivity, quality, and overall performance in organizations. Key aspects of Industrial Engineering and Management include: Process Optimization; System Analysis and Design; Quality Control and Management; Supply Chain Management; Operations Management; Human Factors and Ergonomics; Project Management; Cost Analysis and Financial Management; Decision Analysis.

Overall, Industrial Engineering and Management aims to optimize resources, improve processes, enhance productivity, and ensure the effective and efficient utilization of all elements involved in the production or delivery of goods and services. It is crucial in today's competitive business environment for organizations to stay efficient and competitive.

Production Engineering and Operational Excellence are fields of study and practices that focus on optimizing and improving the manufacturing and production processes within an organization. It combines principles from engineering, management, and operational strategies to enhance productivity, efficiency, quality, safety, and sustainability in the production of goods and services.

Here are the key components of Production Engineering and Operational Excellence: Process Optimization; Operational Excellence; Manufacturing Systems Design; Quality Management; Supply Chain Optimization; Production Planning and Scheduling; Automation and Technology Integration; Health, Safety, and Environmental Management; Cost Management; Performance Measurement and Key Performance Indicators (KPIs); Continuous Improvement and Innovation.

Production Engineering and Operational Excellence are crucial for organizations aiming to stay competitive in the global market by achieving high levels of efficiency, quality, and customer satisfaction while optimizing resources and minimizing waste. It is a multidisciplinary approach that encompasses engineering principles, management strategies, and the effective use of technology to drive operational success.

Meet the Series Editor

Fausto Pedro Garcia Marquez is a Full Professor at UCLM, Spain, with accreditation since 2013. He also holds the position of Honorary Senior Research Fellow at Birmingham University, UK, and serves as a Lecturer at the Postgraduate European Institute. In addition to these roles, Fausto has experience as a Senior Manager at Accenture from 2013 to 2014. He earned his European Ph.D. with the highest distinction. Throughout his career, Fausto has received numerous awards and honors. These include the Nominate Prize (2022), Gran Maestre (2022), Grand Prize (2021), Runner Prize (2020), and Advancement Prize (2018), as well as Runner (2015), Advancement (2013), and Silver (2012) by the International Society of Management Science and Engineering Management (ICMSEM). He was also the recipient of the First International Business Ideas Competition 2017 Award. Fausto's contributions extend to academic publishing, with over 242 papers to his name. Notably, his work has been recognized in journals like "Applied Energy" (Q1, IF 9.746, Best Paper 2020) and "Renewable Energy" (Q1, IF 8.001, Best Paper 2014). His affiliations include the editorial and authorship roles in more than 50 books, with publications through respected publishers such as Elsevier, Springer, Pearson, Mc-GrawHill, IntechOpen, IGI, Marcombo, and AlfaOmega. He has authored over 100 international chapters and holds 6 patents. Fausto serves as the Editor of 5 International Journals and is a Committee Member for more than 70 International Conferences. His research portfolio encompasses being the Principal Investigator in 4 European Projects, 8 National Projects, and participating in over 150 projects involving universities and companies. His areas of expertise and research interests span Artificial Intelligence, Maintenance, Management, Renewable Energy, Transport, Advanced Analytics, and Data Science. Fausto is a recognized Expert in the European Union in AI4People (EISMD) and ESF. He also serves as the Director of www.ingeniumgroup.eu, holds the status of Senior Member at IEEE since 2021, and has been honored as an Honorary Member of the Research Council of the Indian Institute of Finance since 2021. Fausto is also the Committee Chair of The International Society for Management Science and Engineering Management (ISMSEM) since 2020.

Meet the Volume Editors

Germano Lambert-Torres received his Ph.D. in Electrical Engineering from the École Polytechnique de Montreal in 1990. Dr. Lambert-Torres has led research in intelligent systems, maintenance management, and power system operation for many power companies in Brazil and South America. He has had more than 150 R&D projects financed by industries. He has also taught seminars and short courses in many countries worldwide at the Institute of Electrical and Electronics Engineers (IEEE), International Council on Large Electric Systems (CIGRÉ), and International Federation of Automatic Control (IFAC) conferences, universities, and research centers. He has supervised more than 120 MSc and Ph.D. theses and published more than 800 journal and technical conference papers. He is also the author and editor of ten books. He received many awards, including the IEEE IAS Outstanding Educator/Mentor Award in 2023. Dr. Lambert-Torres is a Fellow of the IEEE.

Cláudio Inácio de Almeida Costa obtained a bachelor's degree, master's degree, and a doctorate in Electrical Engineering from the Federal University of Itajuba, Brazil, in 1995, 1997, and 2017, respectively. He has experience in electrical engineering, with an emphasis on electrical power systems, working mainly on control and operation, load forecasting, and intelligent systems (big data, random forest, fuzzy logic, expert system, neural networks, paraconsistent logic). He also has experience in software development, consulting, and R&D projects for many power companies in Brazil. In 1997, he founded Know-How Quality & Technology, which executes computational services for municipalities and cities and for which he is CEO.

Gilberto Capistrano Cunha de Andrade has been an associate researcher at the GNARUS Institute,Itajuba, Brazil since 2021. He has a solid academic background and diverse experience in applied mathematics, statistics, and energy engineering. He earned a Ph.D. in Applied Mathematics/Statistics from the University of Beira Interior (UBI), Portugal, in 2015. He obtained a bachelor's in mathematics from the Faculty of Philosophy, Sciences, and Letters of Itajuba (FAFI), Brazil, in 1999. He later specialized in mathematics and statistics at the Federal University of Lavras (UFLA), Brazil, in 2007 and Applied Geoprocessing at the Federal Institute of Northern Minas Gerais (IFNMG), Brazil, in 2022. He is currently Content Professor at the Education Center of the Federal University of Itajuba (CEDUC/UAB/Unifei-MG).

Contents

Preface

Systems engineering is based on a systemic approach, where all aspects, even those that may initially seem unrelated to the final goal, must be considered. This systemic approach allows the creation of models for the analyzed system, creating a new possibility of analysis of the observed system and its relationship with other existing systems, whether in the real world, the mathematical world, or the computational world.

Assembling models and, most often, several submodels to represent a system allows us to study the various relationships of the system observed both internally and externally. With these multiple submodels, it is possible to see how each part of the system interferes with its final result and how each subsystem triggers (or inhibits) another subsystem. This study is quite interesting, mainly to improve the local performance of the system or even its overall performance. It is not uncommon for an overfit in one subsystem to lead to poorer overall performance of the entire system. Thus, studies can be done to achieve suboptimal adjustments in each subsystem, leading to an optimal overall performance.

In addition, this subsystem approach allows each subsystem to have its model replaced if a better model emerges without needing to change the other models. Another advantage is that this approach allows for a systemic way of modeling a problem through submodels.

Externally, the same is true. Each submodel can be evaluated in its performance (as well as the total system model) by its relationship with the other external systems, both those that provide inputs to the observed system and those that receive the outputs of this system. Each submodel of the observed system can also be studied and evaluated for its performance. It is easier to assess stressful situations, borderline situations, or incomplete data in this situation.

Thus, systems engineering involves a part-whole approach, where the observed system is seen in detail through its subsystems (or submodels). Often, the observed system does not have industrial characteristics (physical systems). That is, it is not composed only of parts, machines, or gears, or processes, rules, and legislation, but it has social characteristics (conceptual systems). For example, a given social system, such as the study of a part of society, whether this division is by region, age, or even through a specific characteristic that groups certain individuals.

The same concepts presented above are still valid for any system. Some use more expressions and mathematical relations, while others use better expressions through linguistic relations. Another exciting element is that systems engineering takes on a multidisciplinary and interdisciplinary character where engineers, mathematicians, economists, and people from computer science, human sciences, and many other areas of knowledge can merge their skills to solve complex systems, dealing with the various phases of the modeling process. In this way, the term "engineering" should be understood in its broadest concept: to conceive something concrete from a set of abstract ideas through scientific methods.

Finally, a modeled system is usually designed to draw conclusions about itself or to interact with other similarly modeled systems (by similar techniques or not). This set of modeled systems creates an increasingly complex system of relationships, allowing for more advanced analyses, where various areas of knowledge can act, obtaining information and adding their knowledge.

So, systems engineering is a field of science that is constantly evolving and has few defined areas. Each case takes advantage of modeling ideas and approaches used in other solutions, allowing past experiences to be used in future modeling. This book presents ideas and implementations of systems engineering techniques so that researchers can develop systemic models as new perspectives and inspirations.

This book is divided into two parts: "Fundamentals" and "Applications". The first part consists of four chapters that address some basic concepts of systems engineering and how systems can be modeled and how specific approaches can be applied. The second part consists of five chapters that present different modeling studies that can be easily transferred to other applications.

Chapter 1, "Tailoring the System Life Cycle", states that each modeling project is unique (although it will always have analogous characteristics of other systems) and must adapt to the modeling life cycle. This adaptation to the life cycle significantly improves the performance of the system modeling, as it allows the use of specific inputs necessary for this modeling period. The chapter presents a structured method that involves modeling the individual life cycles of subsystems. Items such as the temporal sequence of activities, the identification of conduction and support activities, the modeling of the stages of the support life cycle, the modeling of interdependencies and, finally, the gathering of all models composing a single general modeling of the system are observed. An illustrative example of the proposed method is applied to a generic system, which allows its easy reproduction for any system.

Chapter 2, "How to Maintain the Unsustainability of Mobility in the Development of Urban Areas? – Example from Slovakia", describes how to maintain mobility and the traffic system in critical conditions in urban areas. This chapter is interesting because it presents multimodal systems of locomotion, which interact with each other to sometimes complement each other and to sometimes act concurrently. In addition, there are pressure groups, such as environmentalists, activists, urban mobility technicians, and municipal interests, who often confront each other, and all need to participate in the final assembly of the modeling. Initially, the problem of urban circulation could be solved simply by mathematical modeling since it is a question of departure and arrival points, predefined routes, number of vehicles (of all types), and the various modes of transport options. However, this is not that simple. The interests of the groups described above interfere decisively with the final modeling. The chapter presents an example of this approach for the city of Bratislava that can be applied to any other urban center and to problems involving mathematical and behavioral aspects.

Chapter 3, "Enhancing Total Productive Maintenance Strategy for Improved Productivity and Performance", presents an approach to enhance a Total Productive

Maintenance (TPM) strategy. The objective is to improve manufacturing processes' productivity and performance, explaining the existing problems in the TPM implementation process. The systemic approach starts by identifying the main barriers and verifying the main influencing factors that create problems in the modeling and implementation of the process. Then, referrals and ways to solve the various issues are sought. This approach allows companies to perceive the process's bottlenecks and create better productive arrangements. Maintenance processes also receive vital information for the improvement of their practices. The chapter presents an illustrative example of a tube mill that can be applied in any other manufacturing process, even in non-industrial approaches.

Chapter 4, "Assigning Multi-Attribute Targets under Uncertainty with Multiple Suppliers and Stakeholders", presents a scenario where the system should have a model with suppliers and stakeholders. There are several possible situations: one supplier and one stakeholder, multiple suppliers and one stakeholder, one supplier and multiple stakeholders, and multiple suppliers and multiple stakeholders. Each of these scenarios is studied in terms of its characteristics and degrees of complexity. The chapter presents tach scenario's difficulties and solutions. A multi-attribute approach is used so that the best manipulation of the final solution can be made. Some small numerical examples are given to make it easier to understand the approach taken. Many practical problems have this kind of specificity, and this chapter can be the beginning of an approach.

In the second part of the book, Chapter 5, "Advanced System Diagnostics Tools: Innovations and Applications", presents an approach to improving predictive maintenance processes in industrial processes. Initially, it presents computational tools that are augmented by modern artificial intelligence (AI) and machine learning (ML) techniques, in addition to the use of controls such as real-time monitoring of processes, the incorporation of sensors, and Internet of Things (IoT) techniques. Examples are presented to illustrate each part of the approach presented. Next, the chapter presents more futuristic elements that have potential for use in this process, such as quantum computing. In addition, the chapter addresses current aspects of industrial processes such as cybersecurity and data privacy.

Chapter 6, "Innovation in Complex Systems: Identifying Opportunities with Systems Thinking and Powering Change through Design Thinking", discusses the blending of two dominant modeling techniques: systems thinking and design thinking. Systems thinking uses modeling that considers complex relationships between the various subsystems and treats patterns within the observed system. Design Thinking is a more empathetic approach to the problem, often used when there is a lack of data or more tangible inputs, such as in some social issues. The chapter presents these two techniques and shows how they can be unified to model a system.

Chapter 7, "Perspective Chapter: Chip I/O Design Fundamentals, Methodologies and Challenges" presents a practical implementation of modeling a chip I/O system with the design of a configurable I/O driver supporting a wide range of data rates from 0.4 to 8.0 Gbps. The proposed configurable I/O driver can comprise two single-ended voltage mode drivers: a differential voltage mode driver and a differential current mode driver. This latest drive makes it possible to solve problems of electronic signaling incompatibilities between chips, leading to maximum performance in various applications the processor intends. The chapter presents four illustrative diagrams.

Chapter 8, “How Much Is the Cost of Implementing Arithmetic on a Quantum Computer?” presents exciting aspects of quantum computing. Starting with classical computing, the chapter leads the reader to the various aspects of quantum computing, such as quantum bits (qubits), its new challenges, how modeling is done, and how calculations are performed. In addition, the chapter presents examples of real-world applications of quantum computing. Although the costs of a quantum computing are still prohibitive, there are already several advantages of its use, and it is believed that when it is available in cloud processing, its use will be fundamental in any complex modeling.

Chapter 9, “Predictive Software Engineering: Delivering Effective Business Solutions through Custom Software Development”, presents a modeling approach based on Predictive Software Engineering (PSE). This technique, increasingly used by companies and industries, allows them to use specific software to meet their needs and, at the same time, be effective in their solutions and transparent in their conclusions. PSE is based on seven main steps: Meaningful Customer Care, Transparent End-to-End Control, Proven Productivity, Efficient Distributed Teams, Disciplined Agile Delivery Process, Measurable Quality Management and Technical Debt Reduction, and Sound Human Development. The chapter covers all these steps.

Germano Lambert-Torres, Gilberto Capistrano Cunha de Andrade and Cláudio Inácio de Almeida Costa
Gnarus Institute,
Itajuba, Brazil

Section 1

Fundamentals

Chapter 1

Tailoring the System Life Cycle

Thomas Kusel

Abstract

Every system is unique and requires a tailored life cycle to ensure that its development can be managed efficiently. The process of tailoring the system life cycle is generally not well understood and is often poorly executed, which can result in inefficiencies and elevated risks for the project. This chapter defines a tailoring methodology that can assist Systems Engineers and Project Managers to better understand and apply the tailoring process. The tailoring methodology first considers the customization of a life cycle for individual subsystems, followed by a harmonization through the system life cycle. The technological and organizational factors that impact on the customization are identified and their impact on the life cycle is described. The chapter also defines how the life cycle can be optimized. Throughout the chapter, examples are given to illustrate how the methodology can be applied to complex and multi-disciplinary systems.

Keywords: systems engineering, system life cycle management, system life cycle tailoring, systems engineering management, project management

1. Introduction

The perception of how much value Systems Engineering (SE) brings to a project versus its cost and its effort varies widely amongst originations and industries [1]. There are many factors that shape these perceptions, some of which are purely subjective, while others highlight real problems with the application of the SE methodology.

The incorrect application of SE can incur a large amount of unnecessary effort on a project, slowing it down and making it heavy, costly and inefficient. On the other hand, a well-applied, minimum effort SE process that enhances the most important aspects of the project can be invigorating and add massive amounts of value. When starting a project, the question then becomes: "What is the 'minimum effort' that should be applied to this project and what are the 'most important aspects' that the SE process should be focused on?" This is a question about tailoring.

The nature of engineering projects varies widely across industries, organizations, technology domains and application domains. The method of effectively engineering these broad-ranging products varies to such a degree that, on the surface, their similarities are almost unrecognizable. For example, the engineering approach to develop consumer software on a mobile platform seems to be far removed from the engineering approach to building a large new bridge. Yet, these engineering processes all have similar underlying engineering principles and can all benefit from the SE approach, if applied appropriately. Thus, the SE process can look radically different between projects, depending on how it is tailored for the specific project.

A further complication for the application of the SE process is the trend of increasingly multi-disciplinary nature of projects. Many complex systems today involve multiple technology disciplines, such as infrastructure, mechanical, electronics, computing and multiple layers of software. For such systems, the aspect of SE process tailoring is even more important, as it is required that the process is tailored, not only for the system as a whole, but also individually for each subsystem. This is necessary for the SE process to be effective across the different technology domains. At the same time, these subsystem processes need to be harmonized by the overall system process to ensure a coherent approach to the engineering effort.

Although there is some literature describing SE tailoring methodologies [2], the aspect of life cycle tailoring specifically is limited. The growing need for tailoring the system life cycle is not sufficiently supported by publications and training. In general, the SE process and life cycle are documented and taught as a "standard life cycle model". Different life cycle models are sometimes presented as competing design patterns [3] (e.g. V-model vs. agile development), or in other cases, some life cycle designs are labeled as outdated methodologies (e.g. waterfall model). This can leave inexperienced Systems Engineers confused and unsure which model to adopt, or how to apply the "standard life cycle model".

The importance of a well-tailored life cycle model is further emphasized by the need for better integration of Program Management and Systems Engineering [4]. The System Life Cycle is one of the important integration mechanisms. A well-tailored life cycle is a valuable tool to structure both the programmatic and technical structures of a project.

The purpose of this chapter is to provide a practical approach to tailoring the "standard" SE life cycle processes. Specifically, the approach should be applicable for a wide range of application domains and for multi-disciplinary engineering problems. The hope is that such an approach can help Systems Engineers and Project Managers to navigate such complex projects more effectively.

The scope is limited to tailoring the system's technical life cycle. More specifically, the focus is on the development portion of the life cycle from conceptualization until start of operations, which includes the following processes [5]: Stakeholder Requirements Definition; Requirements Analysis; Architectural Design; Implementation; Integration; Verification; Transitioning and Validation.

2. Tailoring methodology's summary

The tailoring methodology is performed in two steps:

a. *Customize the life cycle of individual subsystems*: The major subsystems are identified. For each individual subsystem, the technical and organizational factors that impact on the life cycle are identified. With these factors in mind, the life cycle stages and stage gates are identified. Next, the systems engineering processes are mapped onto the life cycle stages. From this, the essential processes and deliverables for each life cycle stage are described. Finally, the life cycle stages are optimized by considering methods of improving their effectiveness and efficiency.

b. *Harmonizing subsystem life cycles through the system level life cycle design*: After the subsystem life cycles are defined, they are combined using the system life cycle. The various stages of the system life cycle are time-synchronized based on upward and downward dependencies. Additional subsystem stages and milestones are identified, where necessary, to support the system life cycle. Where possible, milestones are combined to reduce the overall workload.

These two steps are described in more detail in Sections 4 and 5, respectively, with examples to illustrate their practical application.

3. Technical and organizational factors that impact the life cycle

3.1 Technical factors

a. *Design-build-test cycle cost*: The cost of going through one complete cycle of design, implementation (build) and verification is a major factor in determining the system's development life cycle. In this case, "cost" refers to the complete cost of the project for performing a full cycle, including the schedule delay. On one end of the spectrum: for technologies with a very low cost of implementation (e.g. software), it is efficient to follow an iterative approach of continuously cycling through the development process, thus benefiting from continuous de-risking and the verification of the implementation. This process places a smaller burden on the design phase, as the design and its implementation can easily be adapted. However, one pitfall of this mindset is a neglect of the up-front requirements analysis and well-thought through, stable architecture, which can result in unnecessary re-work. On the other end of the spectrum: for technologies with a very high cost of implementation (e.g. large-scale civil infrastructure), the implementation is performed only once. This places a large burden on verifying the design upstream, before implementation is started. For technologies that lie somewhere on the spectrum in between these two extremes, the cost and schedule impact of each iteration should be weighed off against the benefits of risk reduction per iteration.

b. *Technology maturity*: Technology maturity has a large impact on the design/development phase of the life cycle [6]. First-of-kind systems that use new technologies often go through one or multiple cycles of the design, prototyping and verification before the concept is sufficiently de-risked to proceed to the final design and implementation phases. On the other hand, mature products with minor changes have a much smaller burden to verify the updated design and go through a much-simplified development cycle, or to design upgrade process before proceeding with full implementation.

c. *Production volume*: Making even small changes to an item that is in high volume production can be very expensive and time consuming. The impact is even more severe if such items are already in operation. For these reasons, systems that are produced in large quantities require a very stringent verification of the design before it is transferred to the production line. This typically involves a design qualification phase where one or a small volume of units are produced and a thorough set of qualification tests are performed, including environmental and endurance testing. Depending on the production volume, an additional design optimization may be needed as part of the industrialization process, followed by a further production line verification.

d. *Rigidity of requirements and boundary conditions*: Some systems have evolving stakeholder requirements or requirements that are difficult to define up-front (for example, commercial software products). For such systems, the requirements are allowed to evolve alongside the architectural design, implementation and verification processes in an iterative development. On the other hand, development of systems that have a well-defined and mature set of requirements and boundary conditions can benefit from a rigorous upstream architectural design stage, which provides a firm foundation for the implementation stage. The degree of rigidity of requirements and boundary conditions is often overlooked when designing software systems.

e. *Operating life, cost and environment*: Systems that have a very long life expectancy, high operating cost or operate in a harsh environment typically require a more rigid design verification phase, including representative prototypes and extensive environmental and durability testing. This requires a well-defined design qualification phase with rigorous verification before proceeding from design to full implementation.

3.2 Organizational factors

a. *Procurement strategy*: If some stages of a system's development are outsourced, the procurement strategy for outsourcing such work impacts on the design of the life cycle. The placement of all major contracts needs to be identified as milestones in the life cycle plan, with a baseline preceding the contract placement. The nature of the contract also needs to be considered in the design of the life cycle. For example, if a large, fixed-price contract is placed for the design and implementation of a subsystem, it is critical that the life cycle process is set up to ensure a stable and fully verified requirements baseline for such a contract. This would require that all the upstream work is completed and reviewed, including a full system architecture verification and subsystem requirements allocation. If, on the other hand, the design portion of the work was contracted as a cost-plus (flexible) contract, it would allow some leniency on the up-front work and enable a concurrent system architecture development.

b. *Industry norms and standards*: Some industries and organizations have well-established, standard life cycle patterns that are followed routinely. This typically occurs where organizations perform repetitive projects that are very similar. This sometimes results in industry-wide standard implementation of life cycle tailoring (e.g. civil engineering). In cases where a subsystem is contracted to an organization that follows such standard life cycle patterns, it is usually efficient to allow such organizations to follow their standard pattern. However, it still requires integration into the system life cycle and the optimization thereof by applying SE processes to complement weaknesses in the standard process.

c. *Regulatory environment*: Regulatory environment can have a significant impact on the system life cycle. Heavier regulation typically requires a more elaborate life cycle with more and stricter stage gates. Examples of heavily regulated environments are nuclear and aviation.

4. Tailoring the subsystem life cycles

4.1 Methodology

Designing the life cycle for a subsystem involves the following activities, while taking into account the technical and organizational factors that shape the system life cycle as described in Section 3:

a. Define the life cycle stages.

b. Allocate organizations to execute the life cycle stages.

c. Based on the organizational allocation, define procurement milestones.

d. Identify stage gates and the baselines that are needed at each gate. The procurements should be aligned with baselines.

e. Identify the applicable SE technical processes across the life cycle stages.

f. Optimize the life cycle for efficiency and effectiveness.

Optimizing the life cycle for efficiency means to eliminate unnecessary activities, or to minimize the work that has limited value to the end user. This could involve the reduction of effort in some of the SE technical processes or the elimination of entire life cycle stages.

Optimizing the life cycle for effectiveness means to focus the most effort on activities that add the most potential value. It is important to identify the SE activities that have the most potential value and to dedicate a larger fraction of the engineering effort on such activities.

4.2 Example 1: production hardware

The first example considers an array of Radio Frequency (RF) Antennas for application in radio astronomy. These antennas are large structures (~30 m) with complex electromechanical actuators. The structure is moved in three orthogonal axes to point the antenna. It has challenging mechanical specifications for structural tolerance and pointing accuracy across a large range of environmental conditions.

This subsystem has the following technical and organizational features that shape the life cycle design:

a. High design-build-test cycle cost (financial and schedule).

b. Low technology maturity: It is a new design and has characteristics and performance requirements that have not been previously proven (i.e. low technology maturity).

c. Moderate production volume: 50 units (i.e. small production volume).

d. Long operating life expectancy: 50 years for structural parts.

e. Harsh operating environment: It will be operated in a remote desert area and controlled remotely.

f. Procurement strategy: Outsource the design to a specialist, then place a separate open tender contract for the production phase to ensure a competitive production cost.

The features described above influence the tailoring of the life cycle as follows:

- Features (b), (d) and (e) drive the life cycle towards a comprehensive up-front requirements analysis and design life cycle, with an emphasis on fully verifying the design for performance and durability before proceeding with production.

- Feature (a) drives the life cycle towards a single design-build-test (prototyping) iteration because the project will likely not be able to afford multiple iterations.

- Features (c) and (f) require an industrialization stage following the design qualification to ensure that the design transfer between the design contractor and production contractor is successful, before commencing with full production.

The resulting life cycle design is shown below (**Figure 1**).

Because this subsystem is quite a complex system in itself, the tailored life cycle follows a typical "V"-style pattern with top-down design and bottom-up integration and verification [7]. But it is also strongly shaped by the two procurement milestones.

Conceptual design: This is an in-house activity that involves defining the stakeholder requirements, performing requirements analysis of the concept driving requirements, identifying candidate conceptual designs, trading off concepts,

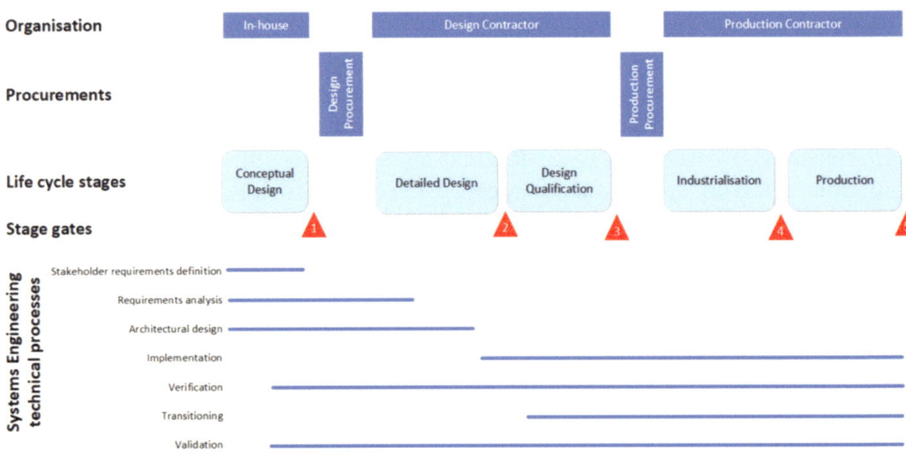

Figure 1.
Production hardware example life cycle design.

designing a preliminary architecture for the chosen concept design, verifying and validating the concept driving requirements and the conceptual design through review processes. The output of this phase is a conceptual design baseline (Baseline 1) that defines the stakeholder requirements, key system requirements and a preliminary design of the chosen design concept. This baseline forms the basis for the procurement of the design contract.

Detailed design: After placement of a design contract, the design contractor performs the identification and analysis of the complete set of requirements and completes a detailed architectural design of the chosen concept, down to leaf component level (i.e. procured component level). Verification and validation of the detailed design is done in the form of a detailed design review, resulting in a detailed design baseline (Baseline 2).

Design qualification: In order to mitigate the significant operational risks of this complex design, the contractor implements a prototype representing the full detailed design. A comprehensive set of qualification testing is performed with the prototype in a representative environment to verify the design. Validation of the prototype is conducted with involvement of the end user. The resulting baseline is a qualified design baseline (Baseline 3) that forms the "blueprint" for the production procurement.

Industrialization: After placement of the production contract, the production contractor sets up the production line for the product, based on the qualified design baseline (blueprint). A small number of units are produced on the production line and installed in the operational environment, followed by a thorough verification of the production process, production documentation and the compliance of the production samples. The resulting production baseline (Baseline 4) consists of the production line, production processes and production documentation.

Production: Following the production baseline gate, the full production process is started. Continuous verification of the produced items is conducted according to the approved production processes. Following the verification of individual production items, they are transitioned into the operational environment, resulting in individual as-built operational baselines for each produced item.

In this example, operational cost optimization is critical. The Radio Frequency (RF) Antennas are expensive and complex systems that operate in a harsh environment over many years. They represent the largest operational cost in the system. There should be an emphasis during the design phase to implement design elements that reduce operational cost (e.g. automatic lubrication systems). There should also be an emphasis during the qualification phase to carefully characterize the reliability of all maintenance significant items, to ensure that the target reliability is achieved before proceeding with production.

4.3 Example 2: software

The second subsystem example considers a software subsystem that is needed to control and monitor a large number of remotely installed Radio Frequency Antennas, in the context of a radio astronomy facility (as defined in Example 1). The software system needs to perform real-time control of all these antennas from a remote location and needs to monitor their health status and performance on a continuous basis, enabling real-time interaction of the array of antennas with end users and with maintainers. This subsystem has the following technical and organizational features that shape the life cycle:

a. Low design-build-test cycle cost. As for most software systems, they can be developed within a virtual test environment and be continuously built and tested.

b. High technology maturity: Well-established control and monitoring protocols are available that can be adapted to this application.

c. High rigidity of requirements and boundary conditions: The external interfaces of the software are well defined. External interfaces are mainly driven by the hardware, which have well-defined interfaces early on in the life cycle. The user interfaces are also fairly easy to determine early in the life cycle.

d. Procurement strategy: Procurement strategy is to develop the software in-house (i.e. no outsourcing).

e. Industry norms and standards: Software development has well-established standards and processes, most often implemented in the form of lean agile development frameworks.

f. Regulatory environment: In this case, there are no strict regulatory requirements. Safety requirements are localized to the antennas, i.e. the control and monitoring software is not safety critical.

The features described above guide the life cycle as follows:

- Features (a), (b), (d), (e) and (f) indicate that a typical iterative software development framework would be an efficient life cycle design.

- Feature (c) indicates that it will be beneficial to perform an up-front requirements definition and architectural design phase, before starting with the iterative implementation.

The resulting life cycle design is shown below (**Figure 2**):

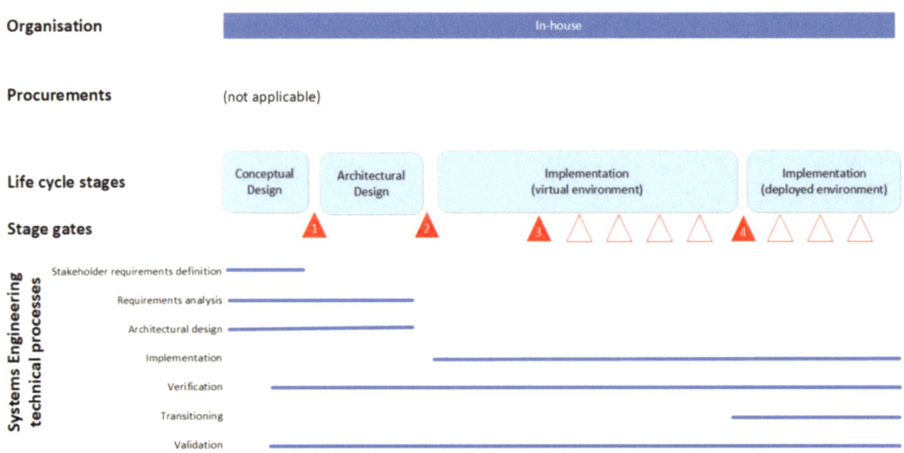

Figure 2.
Software example life cycle design.

The resulting tailored life cycle follows a typical iterative life cycle pattern, but with a clear up-front requirements and architecture design phase before the iterative implementation commences:

Conceptual design: The conceptual design involves defining the stakeholder requirements, performing requirements analysis of the concept driving requirements, identifying candidate conceptual designs, trading off concepts, designing a preliminary architecture for the chosen concept design, verifying and validating the concept driving requirements and the conceptual design through review processes. The output of this phase is a conceptual design baseline (Baseline 1) that defines the stakeholder requirements, key system requirements and a preliminary architectural design of the chosen design concept.

Architectural design: In this phase, a comprehensive requirements analysis is performed. All external interfaces are defined. A complete software architecture is developed based on the requirements and external interfaces. Verification and validation of the requirements and software architecture is done in the form of a detailed design review, resulting in a detailed design baseline (Baseline 2).

Implementation (virtual environment): Software implementation follows a successful detailed design review and is based on the approved detailed architectural design. The software development follows an iterative development with regular planning/reporting cycles using a standard lean agile management framework. The initial implementation is deployed in a virtual environment which includes an integrated test framework where regular software releases (Baselines 3) are verified against the requirements and design. The first software release is a minimum viable product (MVP) that can be integrated into the system for system verification purposes. Validation is performed through regular testing in the virtual environment by end user representatives.

Implementation (deployed environment): When the software is sufficiently mature, software releases (Baselines 4) are deployed iteratively on the operational system based on the needs of the users and their validation feedback.

In this example, an *optimization for efficiency* is achieved by focusing the SE effort on the requirements definition and architectural design phases of the life cycle, where the software processes are often weak. For the implementation phase of the life cycle, the process can mostly rely on the well-established software development methodologies. The SE process can also add value in the definition of software releases during the implementation phase by defining clear, requirements-based targets that align with system level deployment goals.

4.4 Example 3: large-scale computing

The third subsystem example considers a large-scale computing subsystem in the context of a radio astronomy facility. In its final implementation, this subsystem performs a combination of real-time and offline data processing on high volume data received from a radio telescope. The high incoming data rates require real-time processing on a highly parallelized processing architecture that is deployed across many processing nodes in a high-performance computing facility. The purpose of the data processing is to extract the information from the antenna data needed by scientists (astronomers) for research purposes. Because the scientific needs evolve over time, the requirements and the architecture of the software may change over time. This subsystem has the following technical and organizational features that shape the life cycle design:

a. Low design-build-test cycle cost. As for most software systems, it can be developed within a scaled down virtual test environment and can be continuously built and tested, until it is mature to deploy on a full-scale system.

b. Technology maturity: Commercial computing platforms are used for the computing nodes. This technology evolves continuously and the price/performance ratio is continuously declining.

c. Low rigidity of requirements and boundary conditions: Although the interfaces for the incoming data are well defined, the end user interface is not well defined and the user requirements evolve over time.

d. Procurement strategy: Procurement strategy is adopted to develop the software in-house (i.e. no outsourcing) and procure commercially off-the-shelf hardware compute nodes.

e. Industry norms and standards: Software development has well-established standards and processes, most often implemented in the form of iterative development.

The features described above drive the life cycle as follows:

- Features (a), (d) and (e) indicate that a typical iterative, lean agile software development framework is applicable.

- Feature (c) indicates that it is not possible to define a complete set of requirements and detailed design up-front, and that these processes will evolve together with the implementation.

- Feature (b) implies that a late-as-possible procurement strategy for the bulk of the hardware platform would be appropriate.

The resulting life cycle design is illustrated below (**Figure 3**):

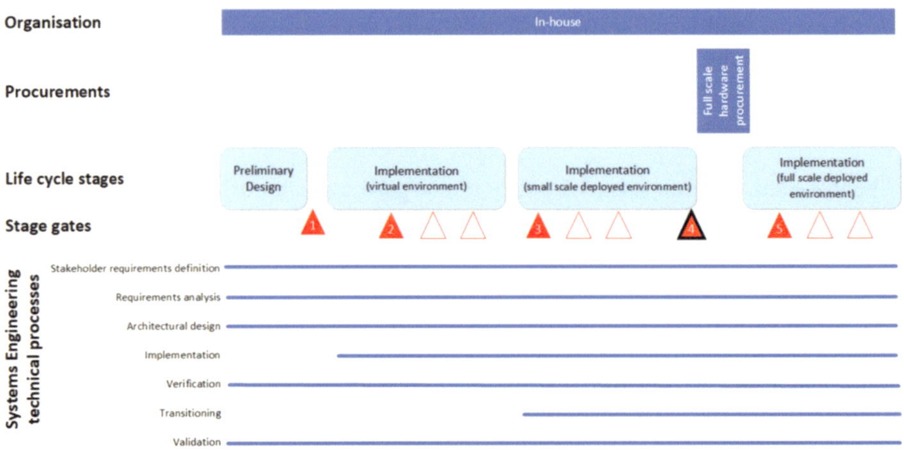

Figure 3.
Large-scale computing example life cycle design.

The resulting tailored life cycle follows a typical iterative life cycle pattern, with limited upstream definition. Deployment of the implementation develops from a virtual environment to a small-scale deployed environment before the full-scale procurement, which is done as late as possible.

Preliminary design: An initial phase of stakeholder requirements identification, requirements analysis and architectural design is performed. The purpose is to identify the hard constraints (e.g. incoming data interface and data rates) and the known stakeholder requirements. The emphasis of this initial requirements analysis is placed on the design driving requirements. From these requirements, an initial architecture is designed, taking into consideration the downstream needs for parallelization. The preliminary design is verified and validated through review processes, resulting in a preliminary design baseline (Baseline 1).

Virtual environment implementation: Software implementation proceeds in a virtual environment, based on the preliminary design. The software development follows an iterative development with regular planning/reporting cycles using a standard lean agile management framework. The initial implementation is deployed in a virtual environment which includes an integrated test framework where regular software releases (Baselines 2) are verified against the evolving requirements and design. The first baseline is a Minimum Viable Product (MVP) that can be integrated into the system for system verification. Validation is performed through regular testing in the virtual environment with end user feedback.

Small-scale deployment implementation: Initial transitioning into the deployed environment is done when the software is sufficiently mature, and the deployed environment is ready. The initial deployment is done on a small scale, with regular software releases (Baselines 3) that are verified against the evolving requirements and design. When the implementation is verified on the small scale and demonstrated to be fully scalable, a formal review is conducted to verify the full-scale architecture design, before procuring the full set of computing hardware. This results in a final design baseline (Baseline 4), which is used as a basis for the hardware procurement.

Full-scale deployment implementation: Procurement of the hardware follows a successful final design review. The software is deployed on the full-scale system, while the iterative software development continues, with regular software releases (Baselines 5). Verification is performed, prior to each new release. Validation occurs through end user operations with the full system.

In this example, an *optimization for efficiency* is achieved by focusing the SE effort on the upstream process of defining the concept driving and design driving requirements and to establish an initial architectural design that satisfies the external hard boundaries. For the implementation phase of the life cycle, the process can mostly rely on well-established software development methodologies. The SE process can also add value in the definition of software releases during the implementation phase by defining clear, requirements-based targets that align with system level deployment goals.

4.5 Example 4: civil infrastructure

The fourth subsystem example considers the building of civil infrastructure required for a typical radio astronomy facility located on a remote green fields site. This includes roads, buildings, electrical reticulation and optical fiber reticulation. This subsystem has the following technical and organizational features that shape the life cycle design:

a. Very high design-build-test cycle cost.

b. High technology maturity.

c. Rigid requirements and boundary conditions.

d. High operating cost and deployment in a remote environment.

e. Procurement strategy: Design and construction supervision to be outsourced to multi-professional engineering consultancy. Construction to be outsourced to one or more construction companies.

f. Industry norms and standards: The civil engineering industry has well-established, standard development and construction processes.

g. Regulatory environment: Significant regulatory constraints (safety, health, environmental, etc.), but well understood by construction contractors.

The features described above drive the life cycle as follows:

- Features (a), (c) and (d) indicate that a thorough requirements definition and detailed design phase are appropriate up-front before proceeding with construction contracting.

- Features (e), (f) and (g) indicate that the design and construction processes will primarily follow the standard civil engineering norms and standards. However, there should be awareness of the weaknesses in these standard processes, and the SE processes should be used to complement such weaknesses.

The resulting tailored life cycle design is illustrated below (**Figure 4**):

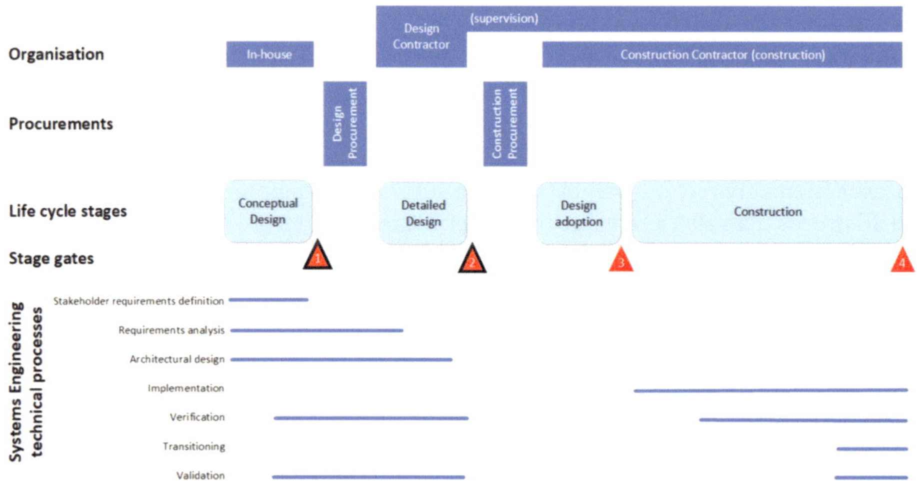

Figure 4.
Civil infrastructure example life cycle design.

The tailored life cycle follows a typical waterfall design pattern, flowing from requirements definition, to design, to implementation and transitioning.

Conceptual design: An in-house conceptual design is done based on the identified stakeholder requirements and the concept driving external interface requirements (i.e. infrastructure needs from other subsystems). Verification and validation of the conceptual design is performed through reviews involving representation from all relevant stakeholders. This results in a conceptual design baseline (Baseline 1), which forms the basis for the procurement of a design contract.

Detailed design: The design contractor performs a complete requirements analysis based on detailed external interface definitions and all relevant regulatory requirements. The design contractor kicks off all the regulatory application processes (which are typically long lead activities). A detailed design is produced based on the requirements analysis. The requirements and design are verified and validated through reviews by representative stakeholders. This results in a detailed design baseline (Baseline 2) that forms the basis for the construction procurement.

Design adoption: The appointed construction contractor performs a design adoption process where the detailed design is transferred from the designers to the construction contractor. Minor adjustments to the design may be needed. The adopted design forms the construction design baseline (Baseline 3).

Construction: During the construction phase, the design contractor performs continuous verification through on-site supervision to ensure that the as-built design complies with the construction design baseline and that as-built documentation is accurate and kept up to date. As the construction work packages are completed, the parts of the system are transitioned to the operational user system and validated by the end users. At the end of the construction phase, a final and comprehensive verification of as-built and operate-to documentation is reviewed, followed by the establishment of the operational and support baseline (Baseline 4).

In this example, efficiency optimization is achieved by focusing the SE effort on the weaknesses of the standard industry civil engineering practices. This includes: (a) the upstream processes in preparation for the placement of the construction contract: requirements definition, requirements analysis, architectural design and verification and validation of the design; (b) the downstream process towards the end of the project: ongoing verification of the construction work and as-built information; transitioning of the system into operations and end user validation.

5. System life cycle

Once all the subsystem life cycles have been tailored, they are combined and harmonized through the system level and stakeholder level life cycle. In this section, we describe the methodology for doing this. We also describe a typical design pattern that is used to combine subsystem, system and stakeholder processes. The chapter concludes with an example by combining the subsystem life cycles that are described in Section 4.

5.1 Methodology

The methodology includes the following activities, which are not necessarily followed in sequence, but typically done concurrently or iteratively:

a. Time-sequence the activities across the life cycles based on dependencies: The Stakeholder, System and Subsystem life cycle stages typically run concurrently, but they are sequenced to ensure that output/input dependencies between activities on the different levels are arranged logically. Typically, the dependencies are top-down (from system to subsystems) during the early stages and bottom-up (subsystems to system) in the later stages. There are sometimes also dependencies between subsystems, as is demonstrated in the example.

b. Identify driving and supporting activities: As the life cycle progresses, some of the activities are driving and some are supporting activities—it is important to distinguish the driving activities clearly, as they will dictate the time sequence of events.

c. Add supporting life cycle stages, if needed: It may be necessary to add life cycle stages to some subsystems to integrate them logically into the system life cycle.

d. Add new milestone gates, if needed: As the interdependencies become clear, it may be necessary to add new milestone gates to subsystem plans to ensure that such dependencies are formalized.

e. Combine milestone gates where possible: There may be opportunities to combine milestones and reviews across the project to reduce the workload on the project team. For example, combining conceptual design reviews of all subsystems with the system level conceptual design review.

5.2 Design pattern

A common design pattern for combining the system and subsystem life cycles is shown below (**Figure 5**).

This pattern, which is based on the well-known "V-model" [7], shows which activities are typically the project driving and supporting activities. Initially, the stakeholder and system level requirements analysis activities drive the development, until the design driving requirements are mature. Following this, the focus is on establishing a mature system architecture supported by subsystem activities

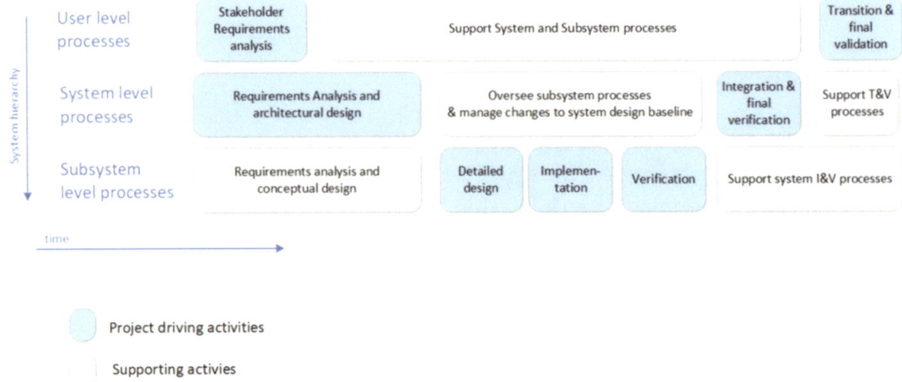

Figure 5.
Design pattern for combining system and subsystem life cycles.

as needed. A baselined system architecture forms the foundation for the subsystem design and implementation. During the subsystem design, implementation and verification activities, the system and stakeholder level activities mostly support subsystem development, while maintaining the system level baseline. When subsystems start to deliver verified products, driving activity moves back to the system level, where system integration and verification typically dictate the activities of the project. Once the system is fully integrated and verified, the focus shifts to transitioning it to the operational environment and final user validation.

5.3 Example application

In this section, the methodology and design pattern described in the previous sections are deployed to define the system life cycle and to combine it with the subsystem life cycles from Section 4.

The combined life cycle is shown below, including the life cycle activities for the system level and all the subsystems. The diagram shows how the life cycle stages and milestones have been arranged temporally to align to critical dependencies between the system level and subsystem level activities (**Figure 6**).

The main features of the combined life cycle are as follows:

a. *System conceptual design*: The purpose of this stage is to define the system concept driving requirements and converge on a single, feasible conceptual system design that can meet all the critical stakeholder requirements. The system conceptual design activity is the main driving activity and is supported by subsystem

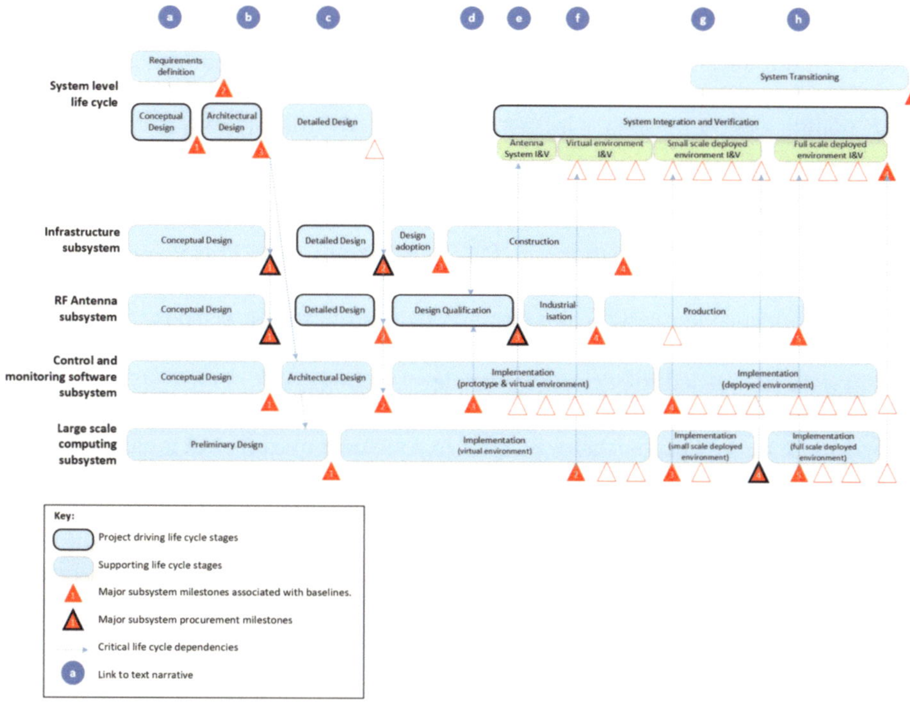

Figure 6.
Example combined life cycle design.

conceptual design development. A number of alternative system design options are generated, which are typically combinations of subsystem concepts. A system level tradeoff study is performed to select the best option. The selection uses a value system that is based on the key stakeholder requirements. This stage is concluded with a conceptual design review, resulting in a conceptual design baseline.

b. *System architectural design*: A detailed system architecture design is developed, based on the conceptual design baseline. A full set of requirements is derived and baselined early on in this process to ensure a complete and stable baseline for the architecture. The development of the system architecture includes a derivation of subsystem requirements and subsystem interfaces. The architecture is verified and validated through a system architectural design review, followed by the establishment of the system architecture design baseline. This baseline constitutes the first critical downward dependency (system to subsystems) in the life cycle. The verified set of subsystem requirements and interfaces represents a firm foundation for the procurement of the infrastructure and RF Antenna design contracts. They also form the basis for the software and computing subsystem design activities.

c. *Detailed design*: At this stage, the subsystem detailed design activities become the driving activities in the project. As the details of the design on subsystem level emerge, the system architecture design baseline is updated and fleshed out, resulting in an updated system architecture design baseline after the subsystem detailed designs have been verified and baselined. The detailed design baseline is an important baseline for the construction procurement of the infrastructure.

d. *Subsystem design qualification*: As explained in Section 4.4.1, the design qualification of the RF Antenna subsystem is an important prototyping activity to verify the as-built design before initiating the production process. For this activity, some subsystem-to-subsystem interdependencies are identified, including the availability of the necessary infrastructure and control software to verify the antenna prototype. During this stage, the system team performs mainly supporting and oversight roles of the subsystem developments.

e. *System integration and verification (I&V) - Antenna system*: This stage represents the first upward dependency (subsystems to system), where subsystem deliverables are needed to establish a system integration, for the purpose of system verification. In this case, a qualified Antenna system and the necessary control software are needed to perform system level testing.

f. *System I&V - Virtual environment*: I&V in a virtual environment is a useful early risk mitigation activity. It allows the early integration and testing of multiple software components before hardware becomes available. This requires the regular deployment in a virtual environment of minimum viable products from the different software teams. A synchronized, iterative cycle of deployment, integration and testing needs to be coordinated between the system team and software teams to set clear goals for each increment and feedback test results.

g. *System I&V – Small-scale deployed environment*: When the hardware is available for a small-scale deployment of the system in the operational environment, and the software development is sufficiently mature, a small-scale deployment of the system may be integrated for deployed system verification. This integration and verification are also performed in a synchronized, iterative cycle. The scale of deployment may also be increased incrementally, depending on the availability of hardware. An early transfer into the user domain may also be done at this stage to de-risk the transfer process. When the small-scale deployment is mature and verified, it provides the confidence to proceed with the procurement for the full-scale computing hardware.

h. *System I&V – Full-scale deployed environment*: Once the full-scale hardware is procured and the production of antennas is completed, the focus shifts to establish a full-scale system for final system verification. This continues in a synchronized, iterative cycle of integration and testing until the functionality is complete and mature. The final successful verification results in a system product baseline, with the final transition to the user domain resulting in an operational readiness baseline.

6. Conclusion

Each project is unique and it is important to tailor the system life cycle to fit the unique characteristics of the system being developed. A well-tailored life cycle can significantly improve the programmatic and technical effectiveness of a project. The system and project characteristics that impact on the design of the life cycle are identified and the impact of the characteristics on the life cycle design is described. A structured method to tailor the system life cycle is presented, which involves first the customization and optimization of individual subsystem life cycles, followed by the synthesis thereof into the system life cycle. The application of the tailoring methodology is demonstrated through the use of various examples, showing how this process can be applied to complex and multi-disciplinary projects.

Acknowledgements

The author acknowledges the South African Radio Astronomy Observatory (SARAO) for their support in producing this publication.

Author details

Thomas Kusel
South African Radio Astronomy Observatory (SARAO), Cape Town, South Africa

*Address all correspondence to: tkusel@sarao.ac.za

References

[1] Ross M, Associates. The value of systems engineering: An industry survey. Journal of Systems Engineering. 2002;**5**(2):105-118

[2] Hill C, Lamiroy B. Tailoring systems engineering processes: A systematic review of literature. Journal of Systems and Software. 2017;**124**:256-269

[3] Beydoun G, Low G. A comparative study of systems engineering life-cycle (SELC) models: Waterfall, spiral, and concurrent engineering. Systems Engineering. 2008;**11**(4):325-346

[4] Rebentisch E. Integrating Program Management and Systems Engineering: Methods, Tools and Organizational Systems for Improving Performance. New Jersey, USA: Wiley; 2017

[5] International Organization for Standardization. Systems and Software Engineering—System Life Cycle Processes. ISO/IEC 15288. Geneva, Switzerland: International Organization for Standardization; 2015

[6] Buchmann T, Koelle R, Schneider K. Incorporating technology maturity into system life cycle models: A systematic literature review. IEEE Transactions on Engineering Management. 2015;**62**(4):472-483

[7] International Council on Systems Engineering. Systems Engineering Handbook: A Guide for System Life Cycle Processes and Activities. 5th ed. New Jersey, USA: Wiley; 2020

Chapter 2

How to Maintain the Unsustainability of Mobility in the Development of Urban Areas? – Example from Slovakia

Tibor Schlosser and Peter Schlosser

Abstract

The chapter discusses the experience of traffic engineering analysis and subsequent transport planning as a comprehensive tool for classifying the suitability of transport design solutions in cities. It will also highlight, on the basis of practical experience, the issue of the interventions of various non-profit organisations, whether for the environment or for alternative modes of transport, who will have to interact professionally and systematically when proposing changes or new solutions for serving the built-up area in cities. The issue mainly affects efforts to develop public transport, the prevailing lack of parking, the suppression of dynamic car traffic, especially in favour of cycling, but all from the point of view of the weakest road transport participant, which are the pedestrians. The public spaces are the relevant issue in a whole line and/or cross section point of view of the road/street. The other important task is the complexity and sustainability of the road/street network in urban areas. The principles focus on the systematic development of sustainable transport in terms of serviceability and accessibility. Distance and time for the functions of land use are crucial parameters and not just some fictional vision of aggressive promotion of one mode of transport at the expense of other modes.

Keywords: traffic engineering, traffic planning, public transport, alternative transport modes, traffic surveys, traffic modelling

1. Introduction

In recent years, the most frequent problems of land use development and its transport infrastructure have most often been related to the requirements of civic and non-profit organisations for project documentation of transport solutions. In the conditions of Slovakia, this is slowly becoming the most serious "problem" for the realisation of any investment in urban areas in the urbanised space for the solutions of transport systems and transport infrastructure. In the conditions of Slovakia, the technical standard for the design of urban roads and its technical regulation have just been updated in 2023 [1, 2]. The traffic-engineering principles that are introduced by the new regulation and the possibilities of complex creative activity of designers will be pointed out.

The main misunderstanding in the development of urban areas on the part of laymen (NGO) is in the basic requirements for the performance of traffic surveys and their processing, and for the area in question in the inevitable need to create a traffic model and its various scenarios as a tool for demonstrating the objectivity and justification of the change in the serviceability of the area. However, this clearly points to fundamental changes in the classification of sustainability of mobility, as it will also need to be demonstrated by measurable parameters in time before and after the changes are made. Examples from existing SUMPs developed in urban settings in Slovakia were used in the analysis. Their content and appropriateness, or shortcomings that develop or violate the principles of comprehensive development of urbanised built-up areas, will be pointed out.

The fundamental problem of "unsustainability" of transport solutions in Slovak cities is the frequent purposeful and unsystematic proposals coming from NGOs and pushing solutions through city councils into the actual implementation of changes in transport infrastructure. Non-professionals always promote imported views as model examples and perspectives or adopt projects from foreign environments into our own territory without knowing the context specific to the particular territory. The aim is to fragment the community living on the land through the many superficial activities of the people who are associated in various associations (non-profit organisations). The aim is usually to promote negativity in the proposed solutions of the authorised engineers with the guidance that we are living badly. Unsystematic solutions hiding behind so-called sustainable mobility solutions are promoted. These solutions are outside the principles of the systematic solutions that are defined in the spatial plans or in their master transport plans. In addition, most of these NGO people do not have the appropriate education for the activities they are criticising and do not want to take responsibility; they just want to dictate subjective demands.

The issue of sustainable mobility has been discussed at several levels in the professional community. We are looking for solutions that respond to sustainable mobility with measurable parameters. These solutions are intended to find answers for the systemic servicing of the built-up area of our cities. The most important thing for us technicians—traffic engineers, planners and urban planners—is to support the requirements for sustainability itself in a measurable way in the meaning of the transport system. This must be addressed through comprehensive transport engineering work and not through subjective declarations in relation to the environment and climate change. Mobility, especially in its expression as a sustainable tool for spatial urban development, rarely refers in its solutions to the essence of people's need, which is accessibility in the quality of the transport service provided. Urban projects and often urban transport service development strategies must address multi-modal spatial arrangements. It is rarely found in analyses and research how the preferred purposeful mobility will be reflected in the future of people's lives in the city. In the literature, we can easily divide research on mobility and its sustainability in areas such as:

1. General or comprehensive solutions of principles, in which transport solutions are included by social instruments of urban development in its multimodality [3–5]

2. Systemic and measurable solutions by improving the quality of service to the built-up area [6–8]

3. Subjective human parameters most often represented by non-systemic and short-term solutions, point-oriented to expedient local benefits preferring some of the "sustainable" parameters

4. Single-purpose preference for one mode of transport—most often cycling, with the expression of the success of solving climate problems without the essence of a comprehensive assessment of the quality of life in cities.

Many references, for example, can be found in city-wide strategies or agglomeration solutions, for example, [9–12]. Some research references address descriptively just the subjective issue of human perception of sustainable mobility with a proposal to translate into nationwide regulations as in Refs. [13–17]. However, most of the literature addresses sustainable mobility mainly by preferring cycling or alternative transport with the manifestation of environmental benefits as the only qualified solutions. However, they do not give an answer to the complexity of the multimodality of the built-up area and their parametric assessments, for example, only by the basic variable, which is accessibility. There is no need to cite these. However, in the process of analysing the issues of negative boundary conditions that we are addressing in Slovakia with the necessary and logically necessary expertise of sustainable mobility, we have few comparable examples in the world. These are complex solutions that go through the analysis of the solved built-up areas by traffic surveys, the design of solutions in several scenarios and, above all, their assessment by modelling and capacity calculations. Nevertheless, in the unprofessional level of political declarations, there are examples of cooperation precisely with non-profit organisations. However, these are not professionally oriented rather they are oriented towards short-term political related decisions. However, such solutions do not give an answer as they are reflected in the outlook in the development of the city and do not provide a concrete classified value of mobility as part of the multimodal transport service in the system of the built-up area. Only one type of transport system usually comes to the fore, which usually constrains and comes at the expense of the others.

2. Urban development and comprehensive transport solutions on the wave of climate change

In our Slovak conditions, the land has historically been inhabited for several generations. The knowledge of nature and the reality of living in an urbanised space may not always be ideal, nor will it ever be, but generations of people have built it up from the opportunities available to them from society at the level of the city, region or state. Unfortunately, there are too few people even in the professions who study, seek and find solutions who know the specific details of the spatial layout of the land use in terms of transport services. The principle is that if we do not know our own historical development of land use with its transport infrastructure and its servicing as a complex service, we easily succumb to any foreign solution that must be immediate to save the environment, where expedient proposals are promoted in order to violate the basic principles of the spatial development system or transport servicing. Indeed, it is often the case, rather the rule, that such solutions are proposed on the basis of the point-by-point and myopically subjective demands of activists. It is not customary that the discussion and understanding of the other side (NGO) can argumentatively justify its demands. Moreover, for the most part, a narrow view without analysis is asserted and, in debates, an inability to sustain an idea and an inability to abandon an idea. Often, point solutions are promoted with a false quick win that literally destroys and annihilates the spatial perception of servicing the area. Most often, the transport infrastructure itself is destroyed.

Experts and responsible planners in the complex activity of land use development, where the main solution is the transport and its service to the land space, must modestly state a situation where it is not proven that the current climatic situation is associated with industrial civilisation and the way of life of people in developed countries. Certain 'experts' claim that climate change is not a natural anomaly, which is precisely why it must be stopped, not treated as a natural thing. The fundamental question is how are the natural conditions that existed sometime around the 1960s to be maintained? It is suggested that this may be linked to the fact that this is a time interval after the present that has not yet faded from the minds of the ideologues of climate alarmism - climate alarmism that has unfortunately already become a political, mainly lobbyist, agenda. This is a fundamental contradiction between individual human experience (even professional experience), which is very short-term, and the nature of climate change, which is very long-term in nature and causes.

In terms of transport and its solution in the built-up area, sustainable mobility solutions are required. Of course, professionals – traffic engineers and urbanists – are not against it. Usually, every comprehensive technical proposal by engineers ends up being "stomped" by activists that car-oriented solutions are being addressed. These need to be "scrapped". In terms of land use and its development, the space we live in and our active actions to change it - both positive and negative - cause us various constraints. However, there are often no explicit reasons to fundamentally change regulations, norms, standards and methodologies. It is often argued only verbally, but not professionally, that these are hardships or potential existential threats. Non-profit organisations are often used as a method to approach problems in an unsystematic way to solve them, and to do so while gaining influence over the people - the citizens, in a way that rises to the level of victimhood. People have to give up something that they consider particularly important, valuable and necessary for their current life in order to satisfy their 'deities'. However, basic measurable parameters are not used. Most often, the basic axiom of 'divinity' is used - in our case, the level of CO_2 in the atmosphere is placed on that pedestal. CO_2 has only a 0.76% influence on the overall warming impact.

3. What to do next for transport planning and sustainable mobility?

In our Slovak conditions, the way and structure of planning - both spatial and transport planning – are also changing, with a significant shift from the "strategic" approaches typical of the 70–80s of the last century to the dominance of market forces in the 90s of the last century lasting to this day. But in urban areas, we are seeing a renewed and developed interest in the role that the planning system has in positioning new land development and in the magnitude of limiting the impact of the car on the urban environment. All around us now are sustainability and smart solutions, but no systematic solution. However, there is a lack of comprehensive Smart Solutions (in the Slovak language, it is specifically about "reasonable" solutions) and New Urbanism in which transport proposals are implemented. These are projects that comprehensively address multimodal demand and accessibility in the urban land use.

There are various forms of space planning that do not rely on the long-term strategic development of the city and are supported by detailed analyses based on databases. Today's methods are simpler, and management is more subtle, often starting with articulating a vision for the city (urban zone). We are now concerned about quality of life and the wide range of employment; social and cultural opportunities significantly influence what a city can offer. This means planning for accessibility for all residents and other newcomers who depend on the city, keeping the local economy vibrant, minimising the negative effects of city life on the environment and offering a wide range of opportunities to participate in the processes of city life. Pausing at this, the professional stated that the effort will be a comprehensive multimodal assessment of the area with a specific proposal. However, we rarely ask in proposals for solutions, what will the demand be and will it change at all? What will be the comprehensive accessibility of people, positively or negatively? Current traffic engineers and planners can act as promoters of change in the city and can play a positive role where progress should be clear, professional and professionally telling in these incremental steps:

- Investigating the basic environment with the tools of theoretical and analytical problems on a specific land use in terms of economy and spatial distribution of functions with an orientation towards mobility. In our perception of the transport service, it involves an analysis of the main modes of transport on the existing transport infrastructure;

- To present the basic theoretical and empirical relationship between transport and the development of the study urban area. This is usually addressed by a transport-capacity assessment of the area;

- Addressing transport in the development of land use by focusing on all modes of transport - rail, road, alternative, pedestrian (the order is not important in this case);

- Joint transport and land development to be provided in overviews and clearly articulated basic assumptions implicitly focusing on the service problems of the area. Everything must be measurably supported so that the technical level of argumentation is not lost. This is most often lacking in the "hard" promotion of alternative modes of transport (cycling and scooters);

- The aim is to determine where the urban land use should move in terms of transport (demand, service, accessibility) from different scientific, cultural and national perspectives, so as to offer capacity transport solutions for the future for the very development of the lives of the generations of people who want to live there.

If we find not only subjective answers (most often only from NGOs) but also solutions from traffic engineers together with urban planners, a richer and deeper discussion can be achieved in solving problems. Here, professionals - engineers, urban planners - must pull the 'activists' with patience into understanding the complexity of the issues of land use. It's not just about their (NGO) point one-point advocacy of their activity (most often pulling in alternative transportation kinds everywhere). Moreover, it should be pointed out that this is the task of professionals and true experts. However, if non-professionals (NGOs) want to take the initiative, they must be held fully accountable! This cannot be solved by NGOs and activists. Third-sector activism will not help on these points. On one fundamental principle because they often solve the situation only on their momentary enthusiasm in a particular project single-mindedly and without responsibility. These people pretend to address sustainable mobility, to "see" climate change, but are never responsible for the actions that are implemented. This is something that every city leadership should take heed of, especially entire city and local councils, which always and everywhere also lack professional knowledge and background and are falsely drawn into action at the level of championing democratic principles. However, the reality is that it is the activities of NGOs that are only minor activists in the spatial arrangement of land use where the traffic problem needs to be addressed comprehensively.

In this view, road (transport) investment will only make a significant difference where the only missing element is a strong economy. That is why in the case of Slovakia, the motorway network is also important (it is a basic infrastructure and not a "superior" one), but somehow, we have not been able to complete it for more than 40 years. On the other hand, there are slumped economies in Western European conurbations where investment in major roads has caused little or no change in the economic outlook. We then take examples from these to Central and Eastern Europe. It is not appropriate, and it is not recommended! However, this does not concern our Slovak specifics at all, even if we already had a completed basic network with all the "former" expressways. Finally, there are cases in which the lack of main road infrastructure does not hinder the performance of the economy. And this is perhaps the hook that the third-sector participants grasp at when discussing each of our new investments. They rarely know how to oppose in a systematic and reasoned way; they put activities on hold and create incredible delays in proceedings.

The draft Multimodal Surface Transportation Efficiency Program also welcomes the activities of amateur people and non-profit organisations into participation, but with one condition, they must have responsibility. And this is unfortunately lacking in Slovak legislation. It needs to be clearly added. No one is against various proposals, but whoever brings "added" value and wants to promote it must have responsibility in the cycle of engineering and administrative activities, even if he is on the "third" side of the issue. It is not the third party (NGO) that sees the issue of the principle of every road and traffic engineer's solution that we must also look at the time and financial side. If the relevant public administrators and planners have the responsibility, then the non-profit, third-party organisations must have it in a democracy!

A similar framework is also useful for evaluating cases from the field of rail transport in urban space, which forms the main theme of multimodal services in urbanised areas. The main problem for NGOs is cars with their infrastructure (which are according to the NGOs, terrible, bad aspects of the solution), while claiming that alternative modes of transport will solve everything. The true and logical solution to transporting large volumes of residents will never be solved by alternative transportation, especially because of accessibility, and rail is a guarantee of quality transportation to serve the built-up area.

Examining the link between rail investment and economic development is less common than it is for road investment. However, according to [18], the same difficulties in teasing out the specific impacts of rail investments as is the case for road infrastructure investments are pointed out. One of the difficulties of summarising conclusions about the impacts of rail transport investments is that it is a more heterogeneous mode of transport than road transport. However, the wide range of rail transports systems with different purposes and different impacts points to opportunities also in our Slovak context. However, the fundamental question is what the transport demand will be in 2050, and will there be sufficient quality of accessibility by different types of transport systems?

The two main problems - transport/environment and transport/economy - are inevitably interlinked. In fact, we should consider the transport + environment + economy triangle together in the land use with mobility, rather than just considering the above pairs separately. Despite our beliefs about cause and effect, transport will be at the centre of economic/environmental discussions whenever the intention of a transport investment is to improve or maintain economic competitiveness. This is not entirely obvious from the "local squabbles" in local contexts around the world because we cannot "focus" it. We just talk, but more often we shout without reasoning (thinking of the third sector in this statement). Ironically, some public transport investments that can generally be regarded as environmental, such as new urban rail systems, will be regarded by some people as environmentally 'intrusive'. All we will do, or rather offer, is a new incentive for the third sector to protest. We need to make it clear to them that part of their thinking must also be simply the size of the demand and the quality of accessibility. For us professionals, these are the basic measurable variables that must be addressed in a multimodal way and not in a purposeful way for one mode of transport.

Despite the complexity of these problems, because we cannot name them, focus them, analyse them and propose solutions in a comprehensive - multimodal - way, it is not difficult to agree with the statement that, for each of the two issues discussed above, we are likely to come closest to a clear result: in the promotion of sustainable mobility packages and smart cities, we need to give priority to the promotion of large-scale integrated public transport systems (and, of course, road transport as well).

This leads to the view that we should also give credit to the all-powerful EU: there are important political, institutional and even cultural factors that determine national attitudes and political views on the importance of transport, the environment and the economy. Why, for example, is political discontinuity so necessary here? Why are we unable to use more enlightened approaches (beware, these are strictly measurable and technically oriented) than any other from the NGO side of the proposals? Why, because of any number of beliefs, do other countries believe in public interventions in general? Answers to these questions might help explain what is happening on land use with transportation and classify the constraints it may be running into. Our problem is how to incorporate this into measurable rules, preferably in standards and technical regulations oriented to transport systems in built-up areas.

In our conditions, this approach cannot be adopted, mainly for this reason: we do not have a complete basic transport infrastructure in Slovakia, neither road, nor rail. However, it is important to point out that there is a need to be aware of this, because even now, and it will certainly grow significantly in the future, the non-profit and third "enviro" sector works with this view. The problem in Central and Eastern Europe is that, after the changes in the early 1990s, we have not yet, and this is more than 30 years ago, resolved the disparities in transport solutions, whether nationally or regionally and especially in cities. But into this unresolved issue dragging on from the past we go pushing "sustainable" solutions not knowing the debt of the lack of transport services to the built-up areas from the past.

It was in the changes in the late 1980s that the barriers were removed. Since then, our cities have succumbed to the freedom of movement of private vehicles, while the transport infrastructure defined in all spatial plans has not been developed at all and has not been developed to the present day. In times long past, the development of our lands was dealt with priority given to public transport services. This development was quickly eliminated by the false freedom of the Western world - the use of private cars. And it continues to this day. There are no resources. But why are there none? It has literally ground to a halt and radically stalled. Transport systems have stopped developing altogether. The structure of subsidies from the state budget has changed; competences have been divided into lower levels - region, district, and municipality - without, of course, financial needs and requirements from central sources. Somehow, it has been forgotten at the level of the state - the departments - that the citizens of the state produce taxes. The economic consequence is the liquidation of the state's natural resources from the primary national resources we had and "thrown away" by privatisation and its monopolisation by global corporations from abroad. Some people were bothered by state monopolisation, but after the "transformation," they do not mind private monopoly nowadays. With this, many states have lost the natural revenue to develop even "transport" solutions.

We are currently overwhelmed by so-called more sustainable transport systems in the urgent timeframe required by climate change. These are serious obstacles to trying to create something new without a system and defining why we have historically over the last 30 years arrived at where we are today. We are deliberately pursuing 'brighter tomorrows' on the principles of liberalism without a gradual and systematic solution based on the past and paying little attention to considerations of forward-looking research on complex sustainable transport solutions, research and visions. We need to know where we are coming from and where we want to go in the development of an urbanised area with good quality transport services and their accessibility. Especially, the third sector of non-profit organisations without responsibility often brings only ad hoc solutions copied from abroad without knowledge of the conditions that are everywhere in the world specific to land use where the quality service of the area is specifically to be addressed. And so, they are significantly starting to drain financial resources. How to systematically define the current situation with the disproportions of our urbanised areas in spatial layout, functional land use, transport service and transport infrastructure? This is a question that activists have not yet, and certainly will not in the future, answer. It is clear that quality land use must address the elements in synergy as:

> *Spatial arrangement with Transport infrastructure and*
> *Functional use of the land use with Transport services.*

These four basic parameters of 'smart growth' in built-up areas or 'sustainable mobility' in urban areas must be addressed simultaneously and by no means in isolation. We return to the problem of supply and demand in the urban area, which is the basic axiom of the servicing of the urban area and must address the following 4 basic parameters in a cross-cutting and interdependent manner: spatial layout - transport infrastructure - functional use of the land use - transport services. However, this has been very slow to emerge in Slovakia so far. Taken together, we should perceive the long-empty coffers of our cities and municipalities and the minimal support for the development of the urbanised areas from the level of the state, which is only interested in building basic infrastructure (the state cannot afford more, after the elimination of the natural state monopolies that brought resources for even these basic needs for development and servicing the built-up areas for the people who live there and pay taxes to the state). The connection of urbanised area units is not always suitable for the benefit of the municipalities, and moreover, it is oriented only to car traffic. These are our mono-functional zones of construction of family houses, or large mono-functional areas of administration and commercial chains all over Slovakia. In its specialism, this is also true of public transport (suburban bus and long-distance public transport). Cosmetic changes in the railway infrastructure in Slovakia are a real response to this. If we want to maintain a certain level of quality of life of our inhabitants with the world, we must objectively state that the current transport policy as well as the policy of land use development of our society (in the hierarchy of state - region - municipality) is not optimal and has large gaps. This is based on the structure of the national economy as well as the strength of the state's own economy, which has been hollowed out by the economic devastation of the turn of the millennium.

4. Sustainable mobility as a smart transport

Future economic analysis of the possibilities of sustainable mobility features will be the right way to extend this research based on our work. We encounter a developing literature on the management of innovations such as 'smart mobility' [19] and 'mobility as a service' [20], but these have not yet incorporated detailed economic analyses with a focus on the long-term development requirements of urbanised areas.

Smart Growth (SG) applies these principles from the street to the regional level [21] (in our conditions in Slovakia, it goes down to the zones). New Urbanism (NU) [22] in our conditions has received increased attention among authorised planners and the general professional public, especially in areas that have experienced conflicts due to uncontrolled development from the early 1990s that persist to this day, in times of privatisation (small or even state). NU's policies are oriented to meet community development needs and improve community - zoning and environmental goals. However, these need to be defined in terms of "measurable quantities" of transport service and not in terms of general declarations often perceived and promoted in the circle of various non-profit and civic associations. This must be done professionally and with responsibility to the public institutions that are legally responsible and accountable for the land use solution. Their first partners are professionals, authorised traffic engineers and urban planners, who design and, through the state, are also guided by legislation, standards and regulations. They must be created and take into account the responsibility by the state uniformly throughout the whole country, and they must apply to everyone - including activists.

It is not up to the third sector to decide. The NGOs take part in participatory meetings and in project discussions and make proposals. This does not mean that their proposals will be incorporated or implemented. The following basic principles can be stated:

- These organisations - civic and non-profit - have no accountability and have not gone through the basic principles of the current democratic rules, which are elections. If they stay at that level (and they want to stay there), they cannot influence decision-making. They can only offer their subjective suggestions, but these may not be considered in administrative processes and for design works as well.

- The members - most often perceived as activists - are not public and state sector employees who are responsible for fulfilling the laws and relevant norms of the state in implementation (if they were, this would be a contradiction of democratic rules). Gradually, we encounter that activists are valid members of councils, or employees of the administration, or local politicians. If this is the case, they can no longer act as activists but as elected representatives with accountability or as valid civil servants subject to the city (region) leadership and the relevant laws.

- Representatives of these (non-profit) organisations are rarely professionals or experts in the respective activity and do not have the relevant education, practice (references) and experience (these cannot be obtained only by criticism and the will to change the current state of affairs because...). Professionalism can only be acquired by taking responsibility for the activities that one carries out oneself; one must design and implement them. It is only after several projects and years, through which one goes with responsibility also for the professions with which one has to collaborate and form a professional team, that one can gradually acquire a truly critical perspective and, above all, always look for solutions (that is, not one solution) and create.

- In most cases of the world, these people are organised by profession in professional chambers, which have their own laws and under which they must work. Here, in any case, the activists do not fit in. Professionals match the quality of their work and their expertise in design with their authorisation in professional chambers working under the law. Only authorised engineers, architects and relevant professionals with professional competence in other professional chambers and organisations under the act may carry out projects and project documentation.

Unfortunately, we often encounter that projects are developed without the appropriate authorisation and professional competence. In addition, people with other authorisations often come forward and verify professional activities that are not theirs by law. They are rarely checked by the public sector. The staff in public institutions also has professional competence, and they certify or require the completion of projects according to the standards and laws in force in the Slovak Republic.

5. Principles of sustainability of transport solutions

In Slovakia, new standards for the design of motorways, roads and urban roads have been developed over the last two years [1, 2]. For the relevant urban areas' development, we have tried to translate the experience of the last three decades into standards and specifications for urban road design and transport services. This section will point out how base tools have been prepared and will already be required especially between the public administration side that approves and permits and the designers and engineers who in turn deliver solutions. The intention of the developers of the urban road design standard was that the 'third sector', which most often confuses and prolongs the processes, should also understand that if it wants to '*add value in sustainable mobility*', it also must meet the basic requirements.

5.1 Extent of land use and principles of transport services

The network of local urban roads for motorised and non-motorised transport shall be compatible with urban and traffic engineering transport planning in accordance with the principles of sustainable development. The requirements for collectors, distributor roads, urban service roads and non-motorised local roads shall be combined with regard to dynamic and parking traffic, taking into account the protection of the environment, agricultural and forest land, nature and cultural monuments taking into account the historical parts of cities.

The principles in the new standards identify threshold values for design elements, but do not offer comprehensive solutions, as a unique approach to the problem to be addressed in the context of specific conditions. These are required to achieve optimal design, given the specificity of each project. These conditions are based on the need to analyse (in measurable conditions):

- Comprehensively the extent of the land use concerned.
- Its spatial layout and functional use of the area in terms of built-up area and population density, and
- The quality of transport infrastructure and transport services determined by the basic transport characteristics: accessibility and traffic load in a multimodal sense.

Depending on the traffic volumes and speeds of individual types of traffic, it is necessary to separate:

- Motor traffic from bicycle traffic and pedestrian traffic.
- Public transport from individual transport.
- Through traffic from origin, destination and intra-urban traffic.
- Cycling (alternative – scooters) traffic from pedestrian traffic.

5.2 Transport network and character of the area

The urban layout is divided into seven spatial perception zones, which are used to classify the area served by the means of transport and the transport system. These were used according to [23]. The servicing of land use is classified by transport infrastructure and transport service. The seven life zones (the occurrence of organisms and human society in the conditions that allow their existence) vary according to the level and volume of physical and social characteristics, providing the tools for the movement of inhabitants from a rural to an urban style of life of society. Urban roads with their complex cross-section parameters are completely a public space. Their base disposition as radial, feeder, distribution, service and others from the technical design description are paired with the technical terminology of roads in design according to urban planning principles (like boulevard, alley, etc.). This way, there is no problem between the traffic engineer and the urban planner (architect) as to how the cross-section of the road should look like in the basic concept.

The most ecologically acceptable solution for transport services in the municipality land use are the modes of transport. The basic split-up division of modal split is: pedestrian, public, alternative and personal vehicular transport. When designing local feeder and service roads, conditions shall be created in such a way that public transport (PT) is preferred and has an independent position in the whole process of creating project documentation and conditions are created for its development in space and time. Traffic management elements shall be designed to create a preference for the passage of PT vehicles, and separate lanes for PT shall be used to the greatest extent possible.

When processing the spatial-planning documents and before starting the design work on the design of a local road (LR) or intersection (also in plural), it is necessary to define the initial traffic conditions determining the scope and quality of the proposed solution. The proposed LR or intersection must meet the anticipated traffic conditions in the municipality and its agglomeration for at least 10+ years and for crossings of state and regional roads for 20+ years after the construction is put into operation.

5.3 Assessment of transport requirements in the area

The calculation of traffic demand is carried out based on knowledge of the current state of transport operation (Level of services - LoS) in the area in question by means of traffic surveys, as well as on the basis of knowledge of the characteristics of the new traffic potential resulting from the planned spatial development of the site in question. The traffic surveys shall be targeted to the area, route or intersection to be addressed. The definition of the size of the study area with adjacent local roads and intersections shall be determined by the municipality.

The municipality determines the extent of the area (example on **Figure 1**) and the network of local roads with intersections based on the valid land use master plan and the upcoming constructions in the vicinity of the addressed area (example on **Figure 2**). For crossings of state and regional roads inside of urban areas, the extent of intersections within the line of that road is also determined by the municipality. Depending on the size of the investment and the number of new road connections to the crossings of state and regional roads, the extent within the line shall be extended to the nearest adjacent intersections.

5.4 Traffic modelling and scenarios

Traffic model scenarios are created from urban disposition of zones to transport network connections (**Figure 3**). The traffic model must be developed for at least the AM and PM peak hour periods. Details are published in [25] (example of creating the traffic model, **Figure 4**):

- The current condition – the year of traffic survey conducted (Example **Figures** 5 and **6**),

Figure 1.
Example of the extent of urban area with developments (Bratislava - Vajnory) [24].

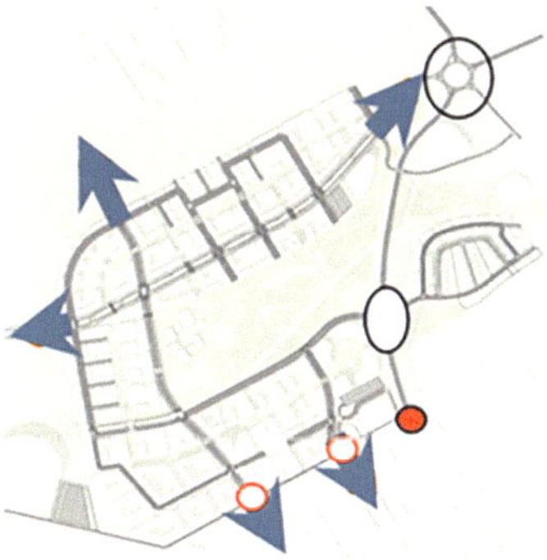

Figure 2.
Definition of the base road network for zone 1 [24].

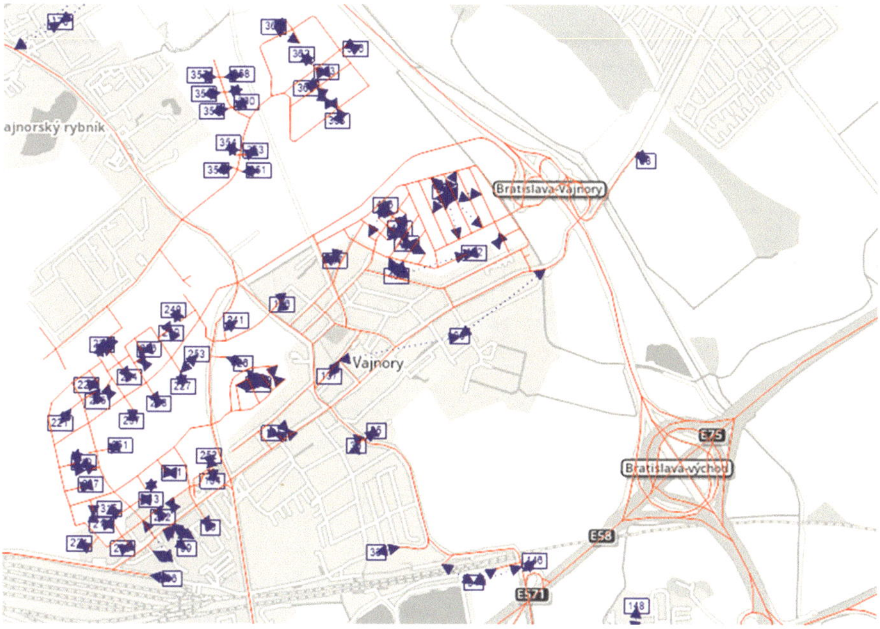

Figure 3.
Example of the distribution of zonal activities on the road infrastructure for the city borough Bratislava – Vajnory (blue arrows are the connectors from the centroid of the zones to the street in model. One zone has a possibility to connect more streets) [24].

- The year of investment opening (day before and day after the opening - this is to demonstrate potential intersection capacity issues), and

- A forward-looking period of +10, +20 years from investment opening as per the function class of local road (example **Figure 7**).

The capacity and throughput traffic of the defined street network and intersections are then assessed against the relevant standards [1] and specifications [27].

The traffic model consists of:

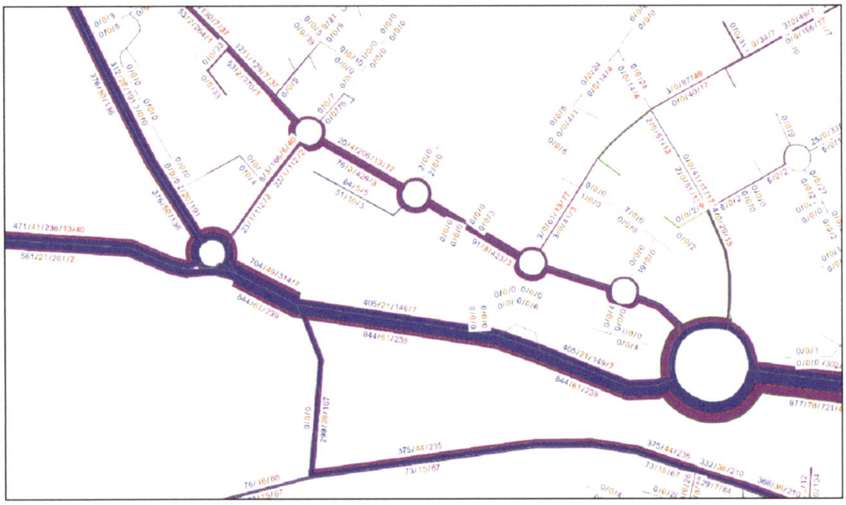

Figure 4.
Example of traffic model output – Scenario perspective of new investment loads – in Bratislava – Bory zone of AM peak hour, 2023. Violet colour - new traffic from investments, blue – Base traffic.

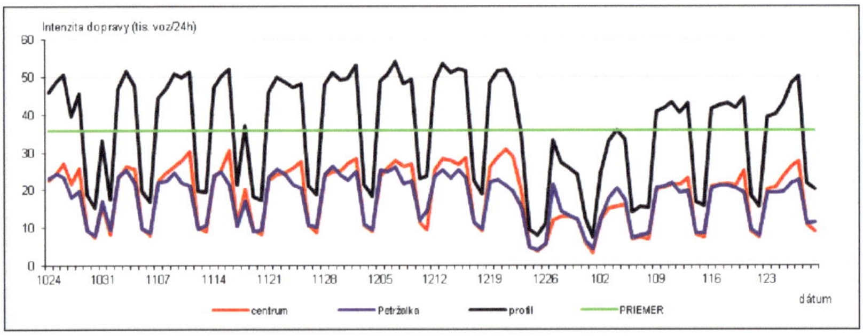

Figure 5.
Example of traffic volume on the street/road for the tracking period of the studied area [26].

- Defining the project and the scope of functions (land use, local roads, calculation of new parking capacities, etc.) – with an example of the traffic model;
- Calculating peak hour volumes of origin, destination and through traffic for each function according to a typical daily pattern (example **Figure** 7);
- Assigning the calculated volumes of origin, destination and through traffic to the proposed road network;
- Traffic-capacity assessment of intersections according to the relevant technical requirements (Example **Figures 8** and **9**) [27];
- Intersectional road sections (between junctions) – their cross-section arrangement;
- Evaluation of the traffic-capacity assessment – time delays and length of queues at individual entrances to the junction.

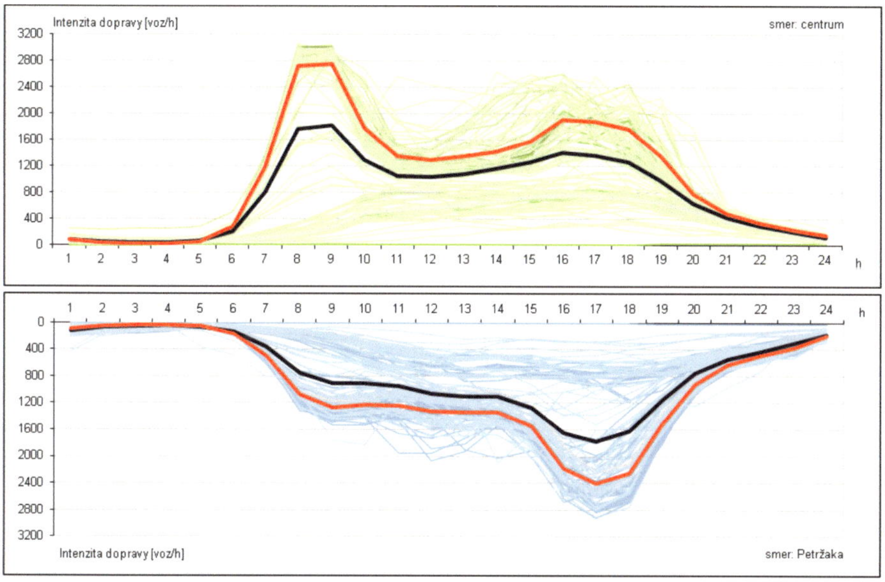

Figure 6.
Example of daily traffic volume on the urban road for the tracking period [26].

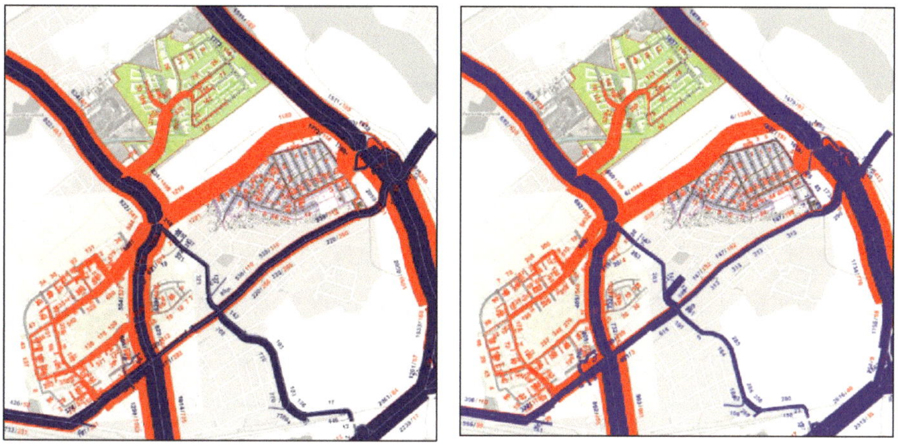

Scenario y. 2040, Peak our traffic AM

Scenario y. 2040, Peak our traffic PM

Figure 7.
Example of the traffic model of the city borough Bratislava - Vajnory with new developments [24].

The results of the traffic modelling will enable to:

- Verify the presented documents and information on the level and quality of dynamic and static traffic congestion by the planned investment;
- Evaluate the proposed technical solutions to optimise the accessibility of the planned investment by all available modes of transport (public, dynamic, cycling, pedestrian, parking) – example of PT Tramway mobility in **Figure 10**;

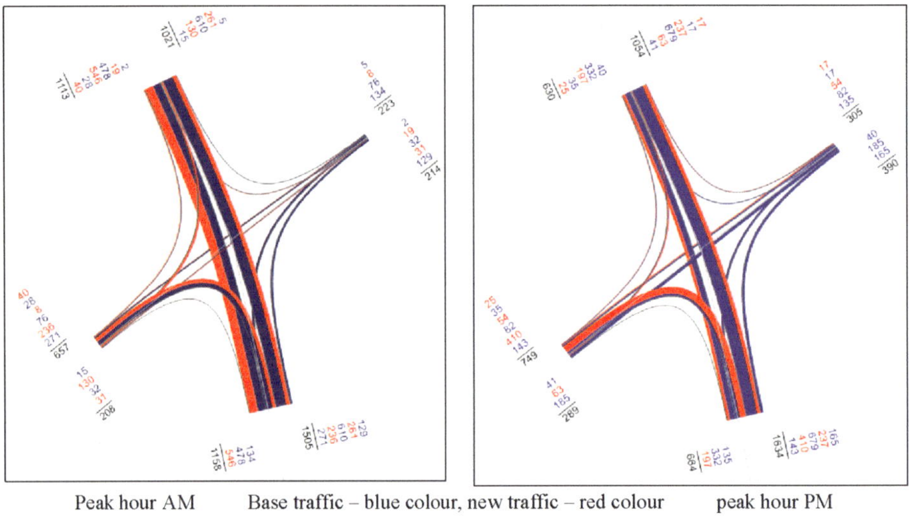

Figure 8.
Graphical example from capacity assessment - evaluated junction output. Peak hour AM Base traffic – Blue colour, new traffic – Red colour peak hour PM.

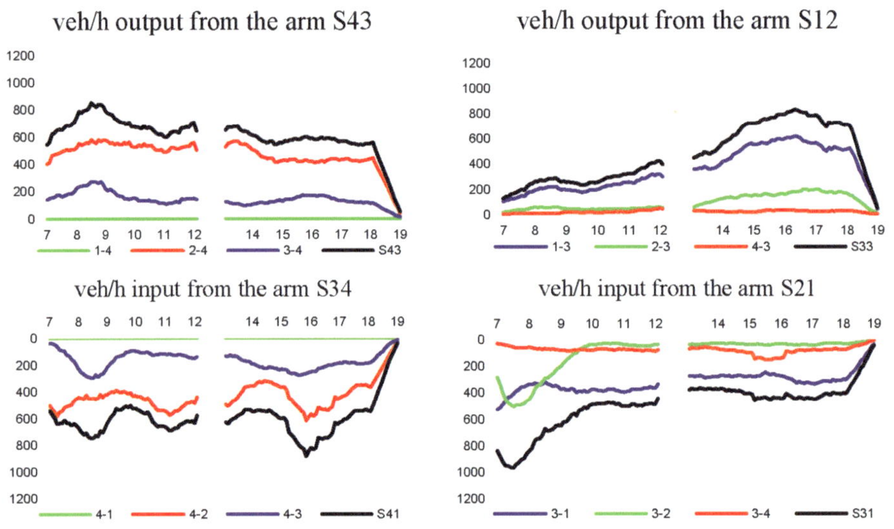

Figure 9.
Example of traffic volume output on the entry and exit of one of the arms of an intersection – Splitting up the traffic flows from each connecting streets on junction [26].

- Analyse the changing of modal split in favour of PT or alternative transport and away from car passenger transport;

- Assess variable scenarios to serve the new area, with proposed stop locations examined by optimising for accessibility in the new developed area (**Figure 11**). An example of an accessibility analysis of bus stops is in [28, 29];

Figure 10.
Example of the resulting public transport solution for new tramway lines around the new Centre in Bratislava [26].

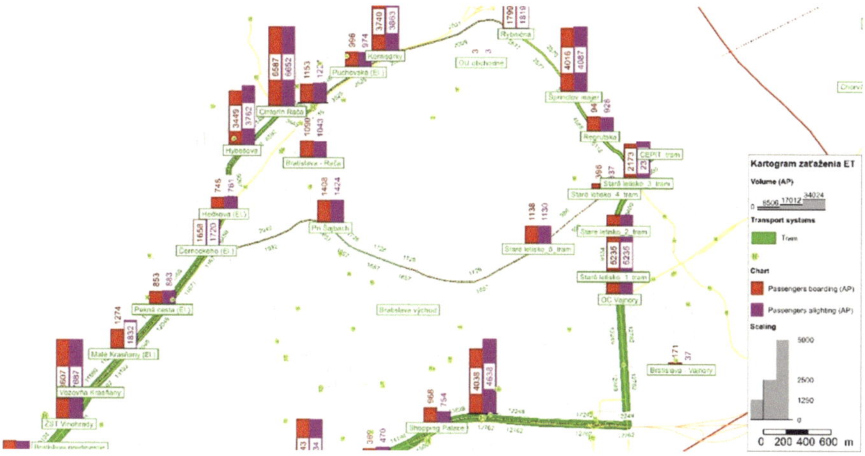

Figure 11.
Example of the output to serve the new area by tramway (according to ***Figure 7****), with proposed stop locations examined by optimising for accessibility in the area.*

- Analyse the technical alignment of the tram line with the values of passenger boardings per 24 h, together with an analysis of the turnover of alighting and boarding passengers at the stops (**Figure 12**). A comprehensive agenda of transport planning and traffic engineering works is presented in [24];
- Propose a solution that does not deteriorate but stabilises and/or improves the existing mobility service of the area for pedestrian, PT and cycling traffic;
- Propose a solution to optimise the provision of accessibility to the area concerned by the planned investment;

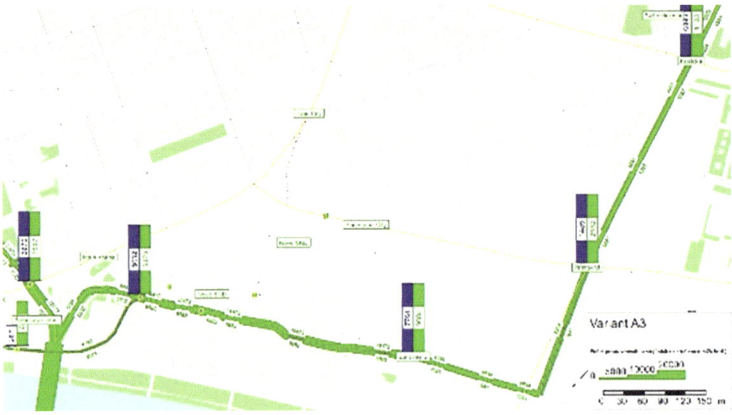

Figure 12.
*Example of the different scenarios of traffic PT model in servicing the area by tramway of the new Centre of Bratislava (passengers/24 h on stops) - detail to **Figure 10**.*

- Propose the phasing of the preparation, the necessary measures and all related activities;
- Assess the degree of congestion on the transport networks caused by the planned investment, and draw the appropriate conclusions.

All these parameters are measurable, not subjective to the will of the "other sides" needs. The empirical experiences are published in [30].

5.5 Traffic capacity assessment and level of service

The traffic capacity assessment (TCA) of investment projects is an integral part of the documentation of their road transport solutions, depending on the type of project documentation. The aim of the capacity assessment is to demonstrate that the solution proposed in the investment project documentation is satisfactory from a capacity point of view. The solution of the other transport segments (public, cycling, pedestrian) is, like the car transport segment, a mandatory part of the investment project documentation. Their scope of assessment is determined by the municipality. The calculation and presentation of the results shall include all relevant values according to [27] and shall be rounded to whole numbers. The form of the outputs is not prescribed. The assessment of the results of the TCA shall be carried out by the administrative body.

In the conditions of Slovakia, the capacity assessment of road transport infrastructure is carried out by calculation. Very popular micro simulations can only be an additional presentation of the calculation, as a graphical 3D output. Very often has it happened that engineers have provided only micro simulations without supporting calculations, which often modify the micro simulations to make the project be perceived positively. This applies to the design of motorways, roads as well as local roads.

The Level of traffic service of the built-up area expressed by the traffic load does not depend on one parameter but on a whole set of parameters of the local road and its facilities in the spatial arrangement of the local road network. It is determined, of course, by the Level of service (LoS), which is the basic criterion of the quality of service offered on the road. It is defined in terms of the quality of vehicle and pedestrian movement offered or the possibility of access to an area served by several modes of transport simultaneously.

The ability of a local road to carry the traffic load (in terms of the number of motor vehicles and other modes of traffic) is assessed at:

- Local road sections between junctions and
- Junctions.

In the street space (public space) for multimodal servicing of land use, it is also required to assess the LoS of pedestrian traffic, cycling and public space at the corners of junctions and at PT stops and PT nodes.

In the development of the strategic documents of the municipality and the city, a functional level of service of the multimodal transport service must be carried out, which establishes the principles of sustainable mobility for the development of the city's transport system. The basic parameters shall be determined for:

- Pedestrians, public transport passengers and cyclists due to longer journeys from origin to destination expressed in terms of dwell time;
- Motor car traffic congestion expressed in terms of saturation of traffic flows on the local road network;
- Inadequate facilities for each mode of transport.

This analysis measures the potential benefits of changes to the transport system for prioritising development in the area. So, in this way, we can define sustainable mobility and smart city solutions for the complexity of citizen's life.

6. Conclusion

The research presented in this chapter is the result of several years of work by planners and traffic engineers who analyse and develop urban areas with multimodal transport according to the above-mentioned projections. Dozens of "battles" are fought each time between experts on the one hand and activists on the other. The role of the professionals is not only as a right hand of developers but also on the side of municipalities as well. The professional work demonstrates, first of all, the measurability of the issue and its evaluation and assessment of several options in urban areas. The automatic criticism of activists against every solution and proposal in any public or private sector project is almost a rule. The fundamental problem are the immeasurable demands of non-profit civic associations and their unsystematic proposals that violate the principles of comprehensive urban development of land use with a multimodal view of servicing. The base task is how to change the tremendous destruction of modal split in Slovakia, which demonstrates exploitation on the side of private car usage. There is often a misunderstanding of the principle of the use of complementary resources, which one chooses in terms of accessibility between the origin and destination of each activity on the urban area. Distance and time are the decisive parameters and not a false will based on the promotion of one mode of transport over another. The systematiic solution lies in the amplification of knowledge, where opportunities for changes in the modal split need to be demonstrated in a measurable way. Only then, with appropriate engineering tools, as outlined in this chapter, can we start to change people's mobility behaviour. The solution appears to be part of integrated transport services, with a focus on public transport with the ability to run regular all-day mobility problems in an urbanised area rather than just covering peak hours. This technical-engineering analysis is a review of 10 years of work by traffic engineers and planners to maintain a clear system approach and solutions with developers and municipalities, which eventually translated into basic principles in a new standard for the design of urban roads in the conditions of Slovakia. However, the open question is how we will be able to transfer this to activists and NGOs, which constantly and always remain without responsibility in the processes. And this is really a problem in the context of the immeasurability of so-called democratic principles in societies almost all over the world.

Author details

Tibor Schlosser[1]* and Peter Schlosser[2]

1 Slovak University of Technology, Bratislava, Slovakia

2 DOTIS Consult. Ltd, Bratislava, Slovakia

*Address all correspondence to: tibor.schlosser@stuba.sk

References

[1] Schlosser T, Schlosser P, Bálint G, et al. STN 73 6110 Projektovanie Miestnych Ciest (Technical Standard: Design of Urban Roads), [In Slovak]. Slovak Office of Standards; 2023

[2] Schlosser T, Schlosser P, Bálint G, et al. TP 73 6110 Projektovanie Miestnych Ciest (Technical Requirements: Design of Urban Roads), [In Slovak]. Ministry of Transport and Construction of Slovak Republic; 2023

[3] Litman T. Mobility management solutions to transport problems around the world. In: Zachariadis T, editor. Cars and Carbon. Dordrecht: Springer; 2012. DOI: 10.1007/978-94-007-2123-4_14

[4] Shiftan Y, Geerlings H, Stead D. Transition towards Sustainable Mobility - the Role of Instruments, Individuals and Institutions, Published by Routledge. London, New York: Taylor Francis; 2016. 430 p. ISBN 978-1-409-42469-7

[5] Yang F, Goldman M, Lagercrantz J. Sustainable Mobility the Chinese Way - Opportunities for European Cooperation and Inspiration [Internet]. 1st ed. Belgium: European Liberal Forum; 2018. 126 p. Available from: https://urn.kb.se/resolve?urn=urn:nbn:se:hig:diva-27901

[6] Bakogiannis E, Kyriakidis C, Siti M, Eleftheriou V. Four stories for sustainable mobility in Greece. In: 3rd Conference on Sustainable Urban Mobility, 3rd CSUM 2016, 26-27 May 2016. Vol. 24. Volos, Greece: Transportation Research Procedia; 2017. pp. 345-353

[7] Sitányiová D, Masarovicová S. Development status of sustainable urban mobility plans in European Union new member states. International Journal of Transport Development and Integration. 2017;**1**(1):16-27. DOI: 10.2495/TDI-V1-N1-16-27

[8] Yadav P, Hasan S, Ojo A, Curry E. The role of open data in driving sustainable mobility in nine smart cities. In: Proceedings of the 25th European Conference on Information Systems (ECIS). Guimarães, Portugal: AIS Electronic Library (AISeL); 2017. pp. 1248-1263. ISBN 978-989-207655

[9] Anastasiadou K, Gavanas N. Enhancing urban public space through appropriate sustainable mobility policies. A multi-criteria analysis approach. Land Use Policy. 2023;**132**. ISSN 0264-8377. DOI: 10.1016/j.landusepol.2023.106765

[10] Tschoerner-Budde C. Sustainable Mobility in Munich - Exploring the Role of Discourse in Policy Change. Springer Fachmedien Wiesbaden; 2019. 249 p. ISBN: 978-3-658-24179-7

[11] Kırdar G, Ardıç İ. S.: A design proposal of integrated smart mobility application for travel behavior change towards sustainable mobility. Civil Engineering and Architecture. 2020;**8**(5):1095-1106. DOI: 10.13189/cea.2020.080536. Available from: http://www.hrpub.org

[12] Krūkle Z, Ernšteins R, Brizga J. Sustainable mobility governance for urban municipalities. In: Latvia: Requirements and Practice of Social-Ecological Systems Approach and Governance Instruments. Sofia: Surveying Geology & Mining Ecology Management (SGEM); 2017. DOI: 10.5593/sgem2017/52

[13] Kovachev A, Slaev A, Nozharova B, Nikolov P, Petrov P. Can Public Participation Contribute to Sustainable Mobility? The Experience of Bulgarian Cities. 2018. Available

from: https://www.researchgate.net/publication/324683897

[14] Maier E. Smart mobility, encouraging sustainable mobility behaviour by designing and implementing policies with citizen involvement. JeDEM - EJournal of EDemocracy and Open Government. 2019;**4**(1):115-141. DOI: 10.29379/jedem.v4i1.110

[15] Whitmarsh L, Gerger Swartling A, Jäger J. Participation of experts and non-experts in a sustainability assessment of mobility. Environmental Policy and Governance. 2009;**19**(4):232-250. DOI: 10.1002/eet.513

[16] Mitran G et al. Sustainable mobility as a result of peoples' awareness on environmental problems generated by transport activity. IOP Conference Series: Materials Science and Engineering. 2019;**568**:012025. DOI: 10.1088/1757-899X/568/1/012025

[17] Lindenau M, Böhler-Baedeker S. Citizen and stakeholder involvement - a precondition for sustainable urban mobility. Transportation Research Procedia. 2014;**4**:347-360. ISSN 2352-1465. DOI: 10.1016/j.trpro.2014.11.026

[18] Breheny M. Transport planning, energy and development: Improving our understanding of the basic relationships, chapter 6. In: Transport and Urban Development. E & FN Spon, an Imprint of Chapman & Hall. London: Taylor & Francis e-Library; 2005. ISBN: 0-203-45806-0

[19] Docherty I, Marsden G, Anable J. The governance of smart mobility. Transportation Research Part A: Policy and Practice. 2018;**115**:114-125. ISSN 0965-8564

[20] Pangbourne K, Mladenović MN, Stead D, Milakis D. Questioning mobility as a service: Unanticipated implications for society and governance. Transportation Research Part A: Policy and Practice. 2020;**131**:35-49. ISSN: 0965-8564

[21] Litman T. Smart Growth, more Efficient Land Use Management. Victoria Transport Policy Institute; 2019

[22] New Urbanism. Available from: http://www.newurbanism.org/newurbanism/principles.html

[23] TRANSECT – Smart Codes. Center for Applied Transect Studies. Available from: https://transect.org/codes.html

[24] Schlosser T, Schlosser P, Bálint G, Korfant M, et al. Technical study of the servicing of the Vajnory Municipal District - traffic study. [In Slovak]. In: Technická štúdia obsluhy územia MČ Vajnory, Dopravná štúdia. STU Bratislava: DOTIS Consult, Ltd. + Department of Transporation Engineering; 2022

[25] Schlosser T, Schlosser P, Zuzulová A, Cápayová S, Bálint G. Traffic planning as a preliminary tool in project documentation for BIM. In Buildings. 2022;**12**(3). ISSN 2075-5309. DOI: 10.3390/buildings12030381

[26] Schlosser T, Schlosser P, Cápayová S, Hodáková D. Complex traffic analysis of new down town of Bratislava for environmental assessment. In: SGEM 2017. 17th International Multidisciplinary Scientific GeoConference. Volume 17. Energy and Clean Technologies: Conference Proceedings. Bulgaria: Albena; 2017. DOI: 10.5593/sgem2017/41/S19.051. ISBN 978-619-7408-06-5

[27] TP 102 Výpočet kapacity pozemných komunikácií. (Technical Requirements: Roads Capacity Calculation). [In Slovak]. Ministry of Transport and Constructions of Slovak Republic; 2016

[28] Schlosser T, Bálint G, Korfant M, Schlosser P. Possible tram-train system in Bratislava line old bridge – Railway Station Petrzalka. In: International Conference on Smart Technologies in Urban Engineering (STUE-2022) - 100th Anniversary of the O. M. Beketov National University of Urban Economy in Kharkiv. Springer; 2022. EasyChair Preprint no. 8226. Available from: https://easychair.org/publications/preprint/ZwQM

[29] Schlosser T, Schlosser P, Cápayová S, Hodáková D, Zuzulová A. Public transport modal Split as a smart growth tool for new city Centre of Bratislava, chapter 6. In: New Approaches in Engineering Research, Volume 4. West Bengal, India & London, United Kingdom: BP International; 2021. pp. 50-78, print ISBN:978-93-5547-355-4. DOI: 10.9734/bpi/nper/v4/13739D

[30] Schlosser T, Schlosser P, Hodáková D, Zuzulová A, Cápayová S. Modelling of capacity and public transport modal Split for new city Centre in Bratislava. In: 4th World Multidisciplinary Civil Engineering-Architecture-Urban Planning Symposium, WMCAUS 2019. Vol. 603. Prague: IOP Conference Series Materials Science and Engineering; 2019. p. 052029. ISSN: 17578981. DOI: 10.1088/1757-899X/603/5/052029

Chapter 3

Enhancing Total Productive Maintenance Strategy for Improved Productivity and Performance

Tamrat Yifter Meles, Alebachew Mengistu Worku, Yewondwosen Gzate Ayalew and Mequanent Ewnetu Belay

Abstract

This study aims to enhance total productive maintenance (TPM) in the pipe industry, specifically within a factory in the Amhara region. The objective is to improve productivity and address challenges faced during TPM implementation. Through identifying obstacles, analyzing factors, and exploring strategies, this research provides insights and recommendations for optimizing maintenance practices. Findings emphasize the need for improved productivity and maintenance strategies, highlighting the importance of a team-based approach. The factory mainly focuses on breakdown and preventive maintenance, lacking a comprehensive TPM approach. The computed overall equipment effectiveness (OEE) stands at 38%, below the standard of 85%. To improve efficiency and reduce downtime, losses limiting equipment effectiveness have been addressed. A total productivity model has been developed, emphasizing the study's significance in enhancing operational performance. Suggested solutions include implementing TPM pillars such as 6S, Autonomous Maintenance, Planned Maintenance, Focused Improvement, Training & Education, Quality Maintenance, Office TPM, and Safety, Health, and Environment, involving management, workers, and maintenance staff. Proposed solutions include training, empowering operators, improving maintenance strategies, and developing a comprehensive TPM implementation plan. Implementing these recommendations will eliminate breakdowns, increase productivity, and improve factory performance.

Keywords: total productivity maintenance, overall equipment effectiveness, autonomous maintenance, productivity, performance, six big losses

1. Introduction

In today's rapidly evolving and fiercely competitive business environment, organizations across industries are constantly striving to achieve optimal productivity and performance. They understand that in order to maintain a competitive edge, it is crucial to ensure the efficient operation of their equipment and machinery while minimizing downtime and maximizing overall equipment effectiveness (OEE) [1, 2]. Total productive maintenance (TPM) has emerged as a comprehensive strategy that aims to achieve these goals by involving all levels of the organization in maintenance activities. However, despite the recognized benefits of TPM, many companies encounter challenges in its successful implementation and fail to realize its full potential [3].

TPM is a holistic approach to maintenance that originated in Japan in the early 1970s as a response to the limitations of traditional reactive maintenance approaches. Such approaches often resulted in unexpected breakdowns, high maintenance costs, and reduced productivity [4]. TPM promotes proactive and preventive maintenance practices, involving all employees in the continuous improvement of equipment and processes. It aims to maximize equipment effectiveness, eliminate losses, and create a culture of ownership and responsibility for maintenance activities [5].

TPM is built on the foundation of eight pillars: autonomous maintenance, planned maintenance, focused improvement, education and training, early equipment management, quality maintenance, office TPM, and safety, health, and environment [6]. These pillars provide a framework for organizations to systematically improve their maintenance practices and drive overall operational excellence [7]. Therefore, enhancing the total productive maintenance (TPM) strategy is of significant importance for organizations aiming to optimize their maintenance practices and achieve improved productivity and performance.

Despite the evident benefits of TPM, its successful implementation remains a challenge for many organizations. The complexities lie in various aspects, such as organizational culture, resistance to change, lack of employee involvement, inadequate training, and ineffective measurement systems. Overcoming these challenges requires a deep understanding of the underlying issues and the development of tailored strategies and approaches [8, 9]. Identifying and addressing these challenges is crucial for organizations seeking to leverage TPM effectively.

To address this problem, a comprehensive examination of the factors influencing TPM implementation is necessary. This research aims to identify the common challenges faced by organizations when implementing TPM and explore effective strategies to overcome these obstacles. By gaining insights into the underlying issues and solutions, organizations can enhance their TPM practices, optimize maintenance activities, and achieve higher levels of productivity and performance.

By addressing the challenges hindering TPM implementation, organizations can enhance equipment effectiveness, reduce downtime, and eliminate losses. This, in turn, will lead to improved productivity and operational performance.

The primary objective of this research is to enhance the total productive maintenance (TPM) strategy for improved productivity and performance by addressing the challenges faced during its implementation. The specific objectives are as follows:

1. Identify the key challenges and obstacles faced by organizations in implementing TPM.

2. Analyze the factors influencing the successful implementation of TPM.

3. Explore best practices and strategies for overcoming the identified challenges.

4. Develop guidelines and recommendations for organizations to enhance their TPM practices.

5. Assess the impact of enhanced TPM implementation on productivity and performance indicators.

By achieving these objectives, this research aims to provide valuable insights and practical recommendations for organizations seeking to optimize their maintenance practices through the effective implementation of TPM. It is anticipated that the findings of this study will contribute to improving productivity, reducing downtime, and enhancing overall performance in various industrial sectors.

2. Literature review

2.1 Definition and principles of TPM

TPM was developed in Japan in the early 1970s and has gained recognition worldwide as an effective approach to maintenance management. Total productive maintenance (TPM) is a comprehensive maintenance strategy that aims to maximize equipment effectiveness, reduce downtime, and empower employees to take ownership of maintenance activities [10].

Cost Savings: Effective TPM implementation can result in cost savings by reducing maintenance expenses, minimizing breakdowns, and optimizing resources. The research will highlight cost-effective strategies and best practices that organizations can adopt to achieve these savings.

Employee Engagement and Ownership: TPM promotes a culture of employee involvement and ownership in maintenance activities. By addressing the challenges impacting employee engagement, organizations can create a motivated and empowered workforce, leading to increased productivity and improved overall performance.

Competitive Advantage: Successful implementation of TPM can provide organizations with a competitive advantage by improving operational efficiency, reducing costs, and enhancing customer satisfaction. The research findings will help organizations gain a competitive edge in their respective industries [11].

2.2 Challenges in TPM implementation

Despite the recognized benefits of TPM, organizations often encounter challenges in its implementation. Several studies have identified common obstacles, including resistance to change, lack of employee involvement, inadequate training, organizational culture, and ineffective measurement systems [12].

2.3 Employee involvement and ownership

Employee involvement is a key principle of TPM, aiming to engage all employees in maintenance activities and foster a culture of ownership. Multiple studies have emphasized the importance of employee involvement for successful TPM implementation. For example, Ranteshwar Singh et al. found that employee involvement positively influenced maintenance practices and overall equipment effectiveness.

They highlighted the need for training and empowerment to foster employee engagement and ownership [7].

2.4 Best practices in TPM implementation

Numerous case studies and research articles have highlighted best practices for successful TPM implementation. These practices include the establishment of cross-functional teams, clear communication of goals and objectives, effective training programs, and the use of visual management techniques. For example, the importance of visual controls and standardized work procedures in TPM implementation enables clarity, consistency, and continuous improvement [13, 14].

2.5 Measurement and performance indicators

An essential aspect of TPM implementation is the measurement of performance indicators to assess its effectiveness. Several studies have proposed various performance indicators, such as overall equipment effectiveness (OEE), mean time between failures (MTBF), and mean time to repair (MTTR). TPM is positively and significantly correlated with low costs (as indicated by higher inventory turns), high quality (as indicated by higher levels of specification compliance), and strong delivery performance (as indicated by a higher percentage of on-time deliveries and faster delivery speeds) [15]. Another study emphasized the significance of OEE as a comprehensive measure of equipment performance and highlighted its use in assessing the impact of TPM implementation.

Additionally, other researchers mentioned that the performance indicators are not defined in isolation but should be the result of a careful analysis of the interaction of the maintenance function with other organizational functions [16]. Therefore, in their study, the performance indicators categorized as measure of equipment performance and cost performance. Under equipment performance, there is failure, breakdown, and overall equipment effectiveness. Under maintenance cost, there are direct maintenance costs, indirect maintenance costs, percent of maintenance cost, and so on [17].

2.6 Integration of TPM with other management systems

TPM can be integrated with other management systems to enhance performance. Integration with Lean Manufacturing principles, often referred to as Lean TPM, combines the waste reduction and continuous improvement aspects of Lean with the equipment-focused approach of TPM. This integration has been shown to improve overall performance and efficiency in various studies [18]. Similarly, the integration of TPM with Six Sigma, another business process strategy, has been employed by companies to enhance manufacturing performance [19, 20].

2.7 Technology enablers for TPM

Advancements in technology have provided new opportunities for TPM implementation. Technologies such as the Internet of Things (IoT), data analytics, and predictive maintenance have been explored to enhance TPM effectiveness. IoT-enabled condition monitoring and real-time data analytics can help predict equipment failures and optimize maintenance activities. For example, Lai et al. discussed the potential of IoT-enabled condition monitoring and real-time data analytics in predicting equipment failures and optimizing maintenance activities [21].

2.8 Case studies of successful TPM implementation

Numerous case studies have demonstrated the successful implementation of TPM in various industries. These case studies provide valuable insights into the challenges faced, strategies employed, and outcomes achieved. For instance, a case study by Almeanazel highlighted the successful implementation of TPM in a manufacturing company, resulting in significant improvements in OEE, reduced downtime, and enhanced employee engagement [22].

2.9 Productivity and maintenance

The main task of production is to manufacture goods/products. However, an efficient maintenance policy significantly influences the production capacity of the machines used for producing these products. Maintenance, therefore, can be considered as an organizational function that operates in parallel with production. While emphasizing that production creates the actual products, the authors also argue that maintenance generates the capacity for production. Total productivity maintenance has a positive impact on the organization's productivity [23, 24].

Measurement of Productivity:

Productivity can be measured either on an aggregate basis or an individual basis. On an aggregate basis, output is compared with the sum of all inputs. This is known as total productivity. On an individual basis, output is compared with one specific input factor, which is referred to as partial productivity or factor productivity:

$$\text{Total productivity index} = \frac{\text{total output}}{\text{total input}} \tag{1}$$

Total production of goods = Labor + Material + Capital + Energy.

2.9.1 Labor productivity

An important aspect of any production setup is achieving the planned quantity of work within a specific time period. Labor productivity depends on how labor is utilized. Factors such as workload availability, materials, working tools, power availability, work efficiency, motivation level, training level, and working conditions (comfortable or poor) can influence labor productivity. Labor can be measured in terms of hours or money:

$$\text{Labor productivity} = \frac{\text{total output}}{\text{labor input}} \tag{2}$$

Labor Productivity (in terms of hours) = Total Quantity Produced/Actual Man-Hours Required.

Labor Productivity (in terms of money) = Total Cost (or Sales Value) of Output Produced/Amount in Rupees Spent on Workers.

2.9.2 Material productivity

In a production system, raw materials are converted into finished products through mechanical or chemical processes. Material productivity plays a crucial role in production costs. It depends on how effectively materials are utilized in the conversion process, considering factors such as rejection percentage, scrap creation, spoilage level, obsolescence, and work wastage:

$$\text{Material productivity} = \frac{\text{total output}}{\text{material input}} \tag{3}$$

2.9.3 Machine productivity

Production systems convert raw materials into finished products using machines and equipment in mechanical or chemical processes. Machine productivity depends on factors such as raw material availability, power availability, workers' skills, and machine layout:

$$\text{Material productivity} = \frac{\text{total output}}{\text{machine input}} \tag{4}$$

2.9.4 Capital productivity

Any production setup requires machines, tools, land, and other assets owned by the organization. Capital is needed for acquiring and maintaining these assets. As a significant amount of capital is invested in these assets, their effective utilization is crucial. Decisions regarding the replacement of fixed assets need to be made. Delaying the replacement of fixed assets may improve the production-to-capital expenditure ratio, but it also increases maintenance costs. Therefore, finding the right balance is necessary. Besides direct expenditures, such as direct material, direct wages, land, buildings, and equipment, a production system incurs indirect expenditures, such as salaries of manpower employed in planning, storekeeping, record keeping, and inspection. Indirect labor is also involved in material movement, good housekeeping, cleaning, etc. Additionally, indirect expenditures are incurred on items such as tools, oils, and lubricants:

$$\text{Capital productivity} = \frac{\text{total output}}{\text{capital input}} \tag{5}$$

3. Research methodology

To achieve the stated objectives, this research will adopt a mixed-methods approach, combining qualitative and quantitative research methods. The research will involve the following steps:

Literature Review: A comprehensive review of existing literature on TPM implementation challenges, best practices, and strategies will be conducted. This will provide a theoretical foundation for the research and help identify research gaps.

3.1 Data collection

3.1.1 Primary data collection

Primary data has been collected through surveys, interviews, and on-site observations. Surveys are being administered to organizations currently implementing TPM, while interviews will be conducted with maintenance managers, employees, and TPM experts. On-site observations will provide a firsthand understanding of the challenges and practices involved in TPM implementation.

3.1.2 Secondary data collection

Additionally, secondary data such as downtime records and information on types of machinery and equipment have been gathered from the case company.

3.2 Data analysis

The collected data will be analyzed using qualitative and quantitative analysis techniques. The qualitative data from interviews and observations will be analyzed thematically to identify recurring patterns and themes. The quantitative data from surveys will be analyzed using statistical tools to determine the significance of factors influencing TPM implementation.

Findings and Recommendations: The research findings will be presented and discussed, highlighting the key challenges faced by organizations in implementing TPM and providing recommendations and strategies for overcoming these challenges. The recommendations will be based on the analysis of best practices and successful case studies.

Below the table to shows that planned and unplanned down time data collection in the factory with reasons for down time. This data used to analysis machine availability to identify running and down time machine for 7 months and to improve overall equipment effectiveness. Let us take some reason for machine failure, shortage of raw material, meeting, break, holyday, tool change, etc. **Tables 1** and **2** illustrate the downtime and machines availability time. **Table 3** presents the machine performance, while **Table 4** displays the monthly production and rejection rate.

3.3 Overall equipment effectiveness

In this section, the overall equipment effectiveness (OEE) has been calculated for Amhara pipe factory based on the data collected by researcher. The OEE analysis considers machine availability, performance rate, and quality rate to measure maintenance and productivity. By identifying the root causes of inefficiencies, the factory can prioritize problem areas and develop a robust maintenance strategy to improve OEE.

3.3.1 Machine availability

Machine availability accounts for downtime losses, including events that halt planned production for a significant amount of time and equipment downtime. Planned downtime, such as holidays or lack of orders, is subtracted from the total operating time to calculate the net operating time:

Net operating time = Total operating time - Planned downtime.

	July		August		September		October		November		December		January	
Reasons	**Planned downtime**	**Unplanned downtime**	**Planned downtime**	**Unplanned downtime**	**Planned downtime**	**Unplanned downtime**	**Planned downtime**	**Unplanned downtime**	**Planned downtime**	**Unplanned downtime**	**Planned downtime**	**Unplanned downtime**	**Planned downtime**	**Unplanned downtime**
Machine down/fault	21	40		20		9		22		18		12		16
Unplanned maintenance		30		18				40		24		30		40
Setup/change over		12		35		17		30		10		10.5		30
Waiting of material		7		30				24		8		20		26
Waiting on quality check		4		15		6				6		5		29
Meetings	10		10				45							
Breaks	45		45			14	45	45					45	
Process warm-up		120		26		20		60		32		31		45
Equipment failures		30		20		8		90		26		60		20
Availability of operator		7		40				10				25		40
Excess capacity	23					288								20
Lack of order		48		7				37		48		24		
Holidays					48		24						24	
Tool damage		52		27		22		59		72		49		34
Total	99	350	55	238	48	384	114	417	0	244	0	266.5	69	300

Table 1.
Down time (planned and unplanned).

Months	Total operating time	Planning downtime (hrs./month)	Net operating time (hrs./month)	Downtime (hrs.)/month	Running time (hrs./month)
July	720	72	570	351	219
August	720	72	593	237	356
September	720	72	555	399	156
October	720	72	579	372	207
November	720	14	661	212	449
December	720	72	603	267.5	335.5
January	720	72	579	300	279
Total	5040	446	4140	2138.5	2001.5

Table 2.
Machines availability time.

Months	Target amount of product	Total actual amount of product	Performance rate (%)
July	281.3	122.7	0.43
August	63359.56	25197.9	0.4
September	226,485	291,036	0.88
October	220,771	220,050	0.9
November	14,950	9095	0.6
December	29,300	20,029	0.68
January	219,025	210,096	0.5
Total	774171.86	775626.6	4.39

Table 3.
Machine performance.

Months	Total input in (m)	Reject/defect in (m)	Output (m)
July	15,203	1570	13,633
August	14,785	1580	13,205
September	15,106	1548	13,558
October	15,798	1917	13,881
November	8211	1427	6784
December	16,054	2017	14,037
January	9209	1560	7649
Total	94,366	11,619	82,747

Table 4.
Monthly production and rejection rate during July to January.

Actual running time is obtained by subtracting downtime losses from the net operating time:

Actual running time = Net operating time - Downtime.

The availability rate is calculated as follows:

$$\text{Availability rate} = \left(\frac{\text{running time}}{\text{net operating time}}\right) * 100$$

Availability rate = (2101.5/4140) * 100 = 50% = (2101.5/4140) *100 = 0.50 = 50%.

3.3.2 Performance rate

Performance rate considers speed loss and minor stoppage loss, which occur when the process operates at less than maximum possible speed. The target output of the machine during running time is compared to the actual output, reflecting speed losses and minor stoppages:

$$\text{Pr} = \left(\frac{\text{Total actual amount of product}}{\text{Target amount of product}}\right) * 100\%$$

PR = (685626.6/774171.86) $*$ 100% = 0.88 = 88%.

Performance losses contribute to the remaining 12% and include minor stoppages and speed loss, such as delays from the production line due to raw material shortages and unnecessary production delays.

3.3.3 Quality rate

Quality rate accounts for quality losses, including produced pieces that do not meet quality standards, require rework or scrap, or have in-process damage. It is calculated by comparing the number of good parts produced to the total number of parts made:

$$\text{Quality ratio } (\%) = \left(\frac{\text{total output} - \text{defect}}{\text{total output}}\right) * 100.$$

QR = [(94366−11619)/94366)] $*$ 100% = 0.87 = 87%.

The quality losses amount to 13% and include defects, measurements, and adjustments. These losses occur during in-progress production and warm-up rejects:

$$\text{OEE} = (\text{MAR} * \text{PR} * \text{QR}\ ) * 100\% = 0.5 * 0.88 * 0.87 = 0.38 = 38\%.$$

The calculated OEE value for the Amhara pipe factory is 38%, which is significantly lower than the world-class OEE. This indicates the presence of various inefficiencies and highlights the need for a robust maintenance strategy. By addressing the six big losses, including breakdowns, setup and adjustment time, idling and minor stoppages, reduced speed, defects and rework, and startup losses, the factory can improve its overall equipment effectiveness and strive toward achieving world-class OEE.

Figure 1 illustrates the comparison between the calculated OEE value for the company and the benchmark for world-class OEE. The existing OEE value for the company is significantly lower, measuring at 38%, while the world-class OEE stands at 85%. This stark difference indicates that the company's equipment efficiency is considerably below the desired standard.

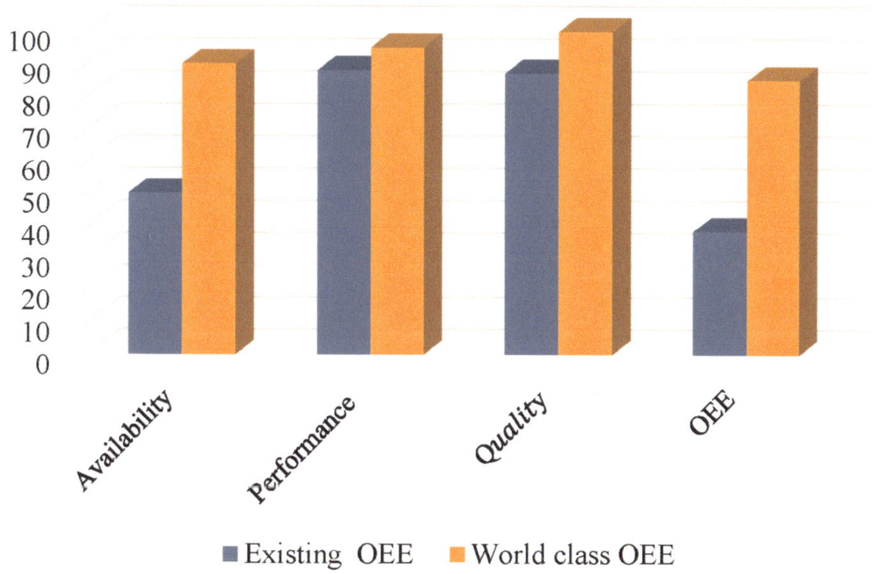

Figure 1.
Comparison of existing OEE with world-class OEE.

To address this challenge, it is crucial for the company to implement a robust maintenance strategy. A strong maintenance strategy can help identify and mitigate the root causes of inefficiencies, such as breakdowns, setup and adjustment time, idle periods, reduced speed, defects and rework, and startup losses. By focusing on these areas, the company can improve its overall equipment effectiveness and work toward achieving higher OEE values comparable to the world-class standard.

Implementing preventive maintenance measures, optimizing production processes, investing in employee training, and adopting modern technologies are some potential approaches that can be part of the maintenance strategy. Continuous monitoring and analysis of OEE metrics will also enable the company to identify areas for improvement and track progress over time.

By prioritizing maintenance efforts and striving for continuous improvement, the company can enhance its equipment availability, performance, and quality rates, thereby increasing its overall OEE and moving closer to achieving world-class standards.

4. Result and discussion

Figure 2 represents a fishbone diagram illustrating the root causes of reduced OEE in the company. It identifies the most significant time-wasting factors contributing to the reduced OEE. Among these factors, the two primary causes are downtime due to machine failure and unexpected machine downtime.

One notable observation is that when such downtime events occur, they are not documented or noted, even though they are the primary reasons for time loss. This lack of documentation hinders the company's ability to analyze and address these issues effectively.

Another key finding from the analysis is that the maintenance staff members are not adequately trained or equipped to address the six big losses that impact OEE.

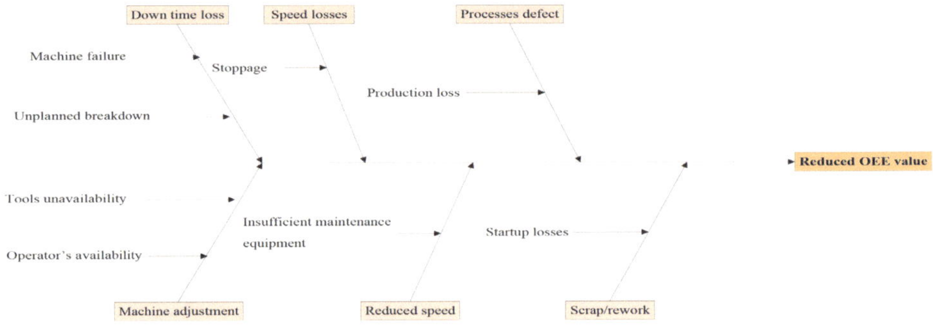

Figure 2.
Cause and effect diagram of reduced OEE value.

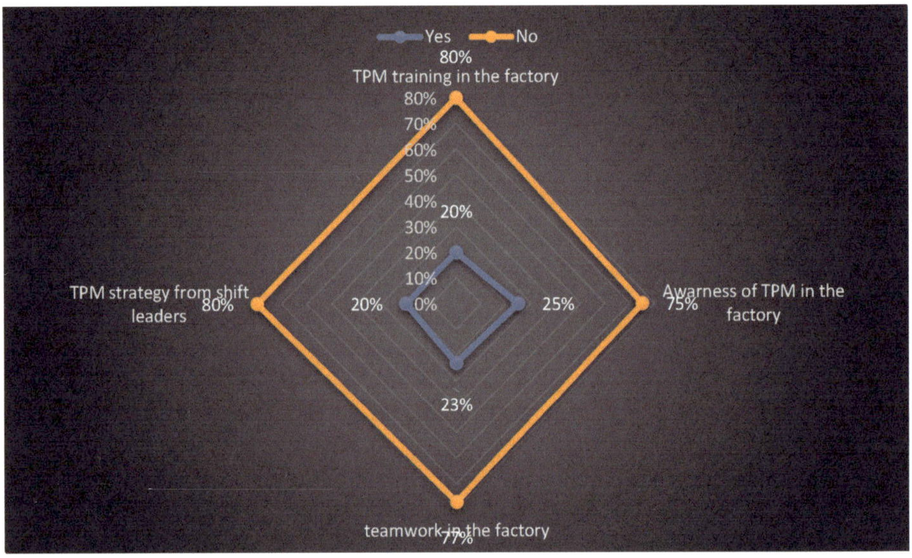

Figure 3.
Percentage distribution of TPM adoptions factory.

Insufficient knowledge and awareness of the importance of identifying and mitigating these losses can further contribute to reduced OEE.

Data from the interwire revealed that in the Amhara pipe factory, 80% of the operators have not received training on TPM, while the remaining 20% have undergone the training. In terms of TPM awareness, 75% of the operators are unaware of TPM, while the remaining 25% have knowledge about it. Furthermore, 77% of the operators in the factory do not work in teams, while the remaining 23% do engage in teamwork. The TPM strategy, as perceived by shift leaders, is also weak. This is because 80% of shift leaders lack awareness and knowledge of the TPM strategy, while the remaining 20% are familiar with it (**Figure 3**).

Figure 4 depicts the root cause analysis conducted to examine the reasons behind the lack of TPM teamwork in the Amhara pipe factory. The analysis revealed several contributing factors. Firstly, one of the main causes identified is the lack of interest and willingness from the management to prioritize TPM teamwork initiatives. This suggests a lack of commitment and support from the management level.

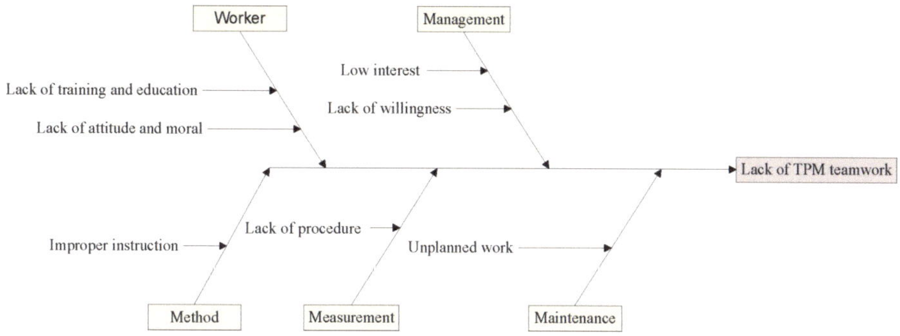

Figure 4.
Cause and effect diagram of lack of TPM teamwork.

Additionally, unplanned maintenance was identified as a key factor. The absence of proper planning and scheduling for maintenance activities can hinder effective teamwork and collaboration among the operators.

Another significant factor is the lack of training and education in TPM principles and practices. When operators are not adequately trained, it impedes their ability to work together efficiently as a team.

The attitude of workers also emerged as a contributing factor. Negative attitudes or resistance toward TPM principles and teamwork can undermine cooperative efforts among the operators.

Furthermore, the analysis highlighted the importance of following appropriate procedures and instructions. Failure to adhere to established procedures and guidelines for TPM teamwork can disrupt the smooth functioning of collaborative efforts.

Lastly, following improper instructions was identified as a factor that hinders TPM teamwork. When operators receive inaccurate or unclear instructions, it can lead to misunderstandings and hinder effective collaboration.

Understanding these root causes is crucial for addressing the lack of TPM teamwork in the Amhara pipe factory. By focusing on these factors, the organization can develop targeted strategies and interventions to foster a collaborative and productive work environment.

To overcome those challenges mentioned above, the following important steps has been has to be followed by the case company.

1. Top Management Commitment: The top management should demonstrate their commitment to TPM by communicating its importance, benefits, and long-term goals to all employees. They should emphasize the need for teamwork and the active involvement of operators and mechanics in maintenance activities.

2. Training and Education: Provide comprehensive training programs on TPM principles, methodologies, and tools for all employees. This training should include autonomous maintenance, which enables operators to perform routine maintenance tasks and identify early signs of problems.

3. Autonomous Maintenance: Encourage operators to take ownership of the machines by conducting regular inspections, cleaning, lubrication, and basic

maintenance tasks. Provide them with the necessary tools, checklists, and guidelines to perform these activities effectively.

4. Planned Maintenance: Establish a structured schedule for planned maintenance activities, including preventive maintenance tasks, equipment inspections, and calibration. Ensure that the maintenance staff and operators work together to plan and execute these activities efficiently.

5. Continuous Improvement: Implement a system for capturing and analyzing data related to equipment performance, breakdowns, and maintenance activities. Encourage employees to identify improvement opportunities and participate in problem-solving activities using techniques such as root cause analysis (RCA).

6. TPM Committees: Form cross-functional teams or TPM committees consisting of representatives from different departments to drive TPM implementation, monitor progress, and address any challenges.

7. Visual Management and Communication: Implement visual management tools, such as visual boards and displays, to share maintenance schedules, equipment status, and performance indicators. Encourage regular communication and feedback between operators, mechanics, and management.

8. Recognition and Rewards: Establish a system to recognize and reward individuals and teams who actively contribute to the successful implementation of TPM and demonstrate improvement in machine reliability and productivity.

By adopting TPM and following these steps, the Amhara pipe factory can create a culture of proactive maintenance, teamwork, and continuous improvement. This approach will help eliminate machine failures, reduce downtime, improve productivity, and effectively address the six big losses, ultimately leading to improved overall equipment effectiveness and organizational success.

4.1 TPM implementation strategy/plan

The implementation of total productive maintenance (TPM) requires a well-defined strategy or plan to ensure its successful adoption and integration into the organization's processes. Based on the Japanese Institute of Plant Maintenance (JIPM) TPM implementation process, along with the eight pillars method, the following steps can be considered for developing a TPM implementation strategy (**Table 5**).

It is important to note that this is a generalized framework, and the specific activities and timeline may vary based on the needs and characteristics of the Amhara pipe factory. The implementation strategy should be tailored to fit the factory's requirements and should involve iterative processes, frequent evaluation, and teamwork to ensure the successful implementation of TPM.

4.2 Total productivity maintenance model

The proposed total productive maintenance (TPM) model consists of four major phases, with each phase having specific activities and responsibilities. The model

Type	Phase 1	Phase 2	Phase 3	Phase 4
Education and training	Conduct basic training on cleaning practices to create awareness and understanding among employees.	Provide team training to enhance collaboration and teamwork among employees involved in TPM implementation.	Focus on training for operation and maintenance, ensuring that employees have the necessary skills and knowledge to perform their tasks effectively.	Analyze and evaluate the training programs to identify areas for improvement and make necessary adjustments.
Preventive maintenance	Reduce the variability of machine lifespan and restore neglected equipment through appropriate maintenance practices	Implement measures to keep the length of machine lifespan and address design weaknesses to prevent unexpected breakdowns.	Develop a schedule for machine lifespan repair and identify the causes of deterioration to proactively address them.	Utilize diagnostic techniques and conduct technical analysis/studies to predict the lifespan of machines and identify potential major breakdowns.
Autonomous maintenance	Basic training, address sources of problem, set cleaning & lubrication	Conduct general inspections on equipment to ensure their proper functioning.	Shift toward autonomous inspection, where operators take responsibility for inspecting and maintaining their equipment.	Establish an organized order line and promote autonomous activities within the organization to ensure equipment reliability and maintenance effectiveness.

Table 5.
Major function involved the development of TPM strategy.

emphasizes the active involvement of top-level managers in driving TPM implementation. Here is a breakdown of the four phases:

Phase 1 (1–2 months):

- Establish TPM strategy: Develop a strategic plan outlining the objectives, targets, and timelines for TPM implementation.
- Assess equipment and necessary materials: Evaluate the condition and performance of equipment and identify any deficiencies or requirements.
- Report findings to top-level managers: Present the assessment results to the management team, highlighting areas that need improvement or attention.
- Develop improvement strategy: Formulate a detailed plan for addressing the identified issues and improving equipment reliability and performance.
- Incorporate maintenance policy: Review and update the maintenance policy to align with TPM principles and objectives.

Phase 2 (3 months):

- Survey employee baseline: Conduct a survey to assess the current knowledge, skills, and attitudes of employees regarding maintenance practices and TPM.
- Establish planned maintenance: Develop a structured system for planned maintenance activities, including preventive maintenance tasks and scheduled inspections.
- Restore targeted equipment: Address the maintenance needs of specific equipment identified in the previous phases.
- Develop material planning feedback: Establish a feedback mechanism for material planning, ensuring timely availability of required resources for maintenance activities.
- Include maintenance improvement feedback: Create a system to capture feedback and suggestions for improving maintenance processes and procedures.

Phase 3 (3–5 months):

- Restore all target equipment: Complete the restoration and maintenance activities for all identified target equipment.
- Roll out maintenance improvement system: Implement an improved maintenance system based on TPM principles, including standardized procedures and workflows.
- Conduct gap analysis: Analyze the gaps between current maintenance practices and TPM objectives, and develop plans to bridge those gaps.
- Reduce the six losses: Address the six major losses in manufacturing processes, which include breakdowns, speed losses, startup and adjustment issues, idling and minor stoppages, quality defects and rework, and startup yield.
- Install planned maintenance: Ensure that the planned maintenance system is fully operational and integrated into daily maintenance activities.

Phase 4 (2–3 months):

- Meticulously identify and reduce big losses: Continuously monitor and identify areas of significant losses and implement strategies to minimize or eliminate them.
- Establish a continuous maintenance system: Create a culture of ongoing maintenance and improvement, with regular inspections, monitoring, and adjustments.

- Integrate maintenance with productivity: Foster a close relationship between maintenance activities and overall productivity goals of the factory.

- Conduct frequent follow-up: Regularly review progress, address any challenges, and provide necessary support to ensure the sustainability of TPM practices.

- Conduct training and education: Provide comprehensive training and education programs for employees to enhance their knowledge and skills in TPM and maintenance practices.

- Establish TPM activities: Formalize the TPM practices and activities within the factory, making them an integral part of daily operations.

- Achieve TPM success: Through the completion of the previous phases' activities, reach a state where TPM practices are fully implemented and contribute to improved productivity and equipment reliability.

By following this proposed TPM model, Amhara pipe factory can implement TPM effectively, improve maintenance practices, reduce losses, and integrate maintenance activities with overall productivity goals. However, it is important to adapt and customize the model to suit the specific needs and circumstances of the factory (**Figure** 5).

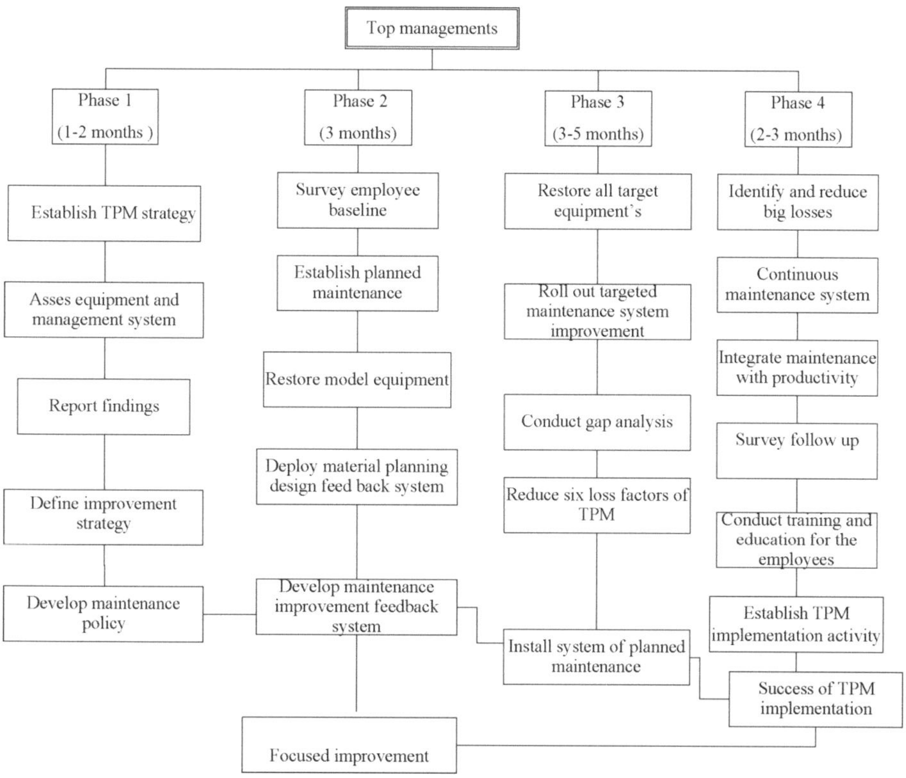

Figure 5.
Proposed TPM model.

5. Conclusion

In conclusion, this research focuses on enhancing the Total Productive Maintenance (TPM) strategy in the pipe industry in the Amhara region, specifically for the pipe factory. The objective is to improve productivity and performance by addressing the challenges faced during TPM implementation. By identifying key obstacles, analyzing influencing factors, and exploring effective strategies, this research provides valuable insights and practical recommendations for organizations aiming to optimize their maintenance practices.

The findings highlight the need for improved productivity and maintenance strategies, as well as a shift toward a team-based approach in the factory. The total productive maintenance strategy should encompass sufficient resources to ensure the effective availability of machines, leading to increased productivity. The research reveals that the maintenance activities in the factory are primarily focused on breakdown and preventive maintenance rather than a comprehensive Total Productive Maintenance approach.

The computed overall equipment effectiveness (OEE) for the case company is 38%, with a machine availability rate of 50%, performance rate of 88%, and quality rate of 87%. These figures indicate that the factory falls significantly below the world-class standard OEE of 85%. To improve machine breakdowns and reduce downtime, it is crucial to identify the causes of loss that limit overall equipment efficiency.

The research suggests implementing the pillars of TPM, including 6S, Autonomous Maintenance, Planned Maintenance, Focused Improvement, Training & Education, Quality Maintenance, Office TPM, and Safety, Health, and Environment. These pillars should be embraced by top managers, shop floor workers, and maintenance staff in the factory to effectively implement Total Productive Maintenance.

Proposed solutions for the factory include providing adequate training for all employees, including maintenance workers and machine operators, empowering operators to carry out small maintenance activities, improving standardized maintenance strategies, and developing a comprehensive TPM implementation plan.

By implementing these recommendations, the factory can eliminate machine breakdowns, increase productivity, and extend the lifespan of their machines. The adoption of a Total Productive Maintenance approach will lead to improved overall equipment effectiveness and reduced downtime, ultimately enhancing the factory's performance and productivity levels.

Author details

Tamrat Yifter Meles[1*], Alebachew Mengistu Worku[2], Yewondwosen Gzate Ayalew[1,3] and Mequanent Ewnetu Belay[1,4]

1 Department of Industrial Engineering, Institute of Technology, University of Gondar, Gondar, Ethiopia

2 Department of Industrial Engineering, Institute Technology, Bahir Dar University, Bahir Dar, Ethiopia

3 Department of Mechanical Engineering, Institute of Technology, University of Gondar, Gondar, Ethiopia

4 Department of Industrial Engineering, University of Gondar, Gondar, Ethiopia

*Address all correspondence to: tamratyifter@gmail.com

References

[1] Zammori F, Braglia M, Frosolini M. Stochastic overall equipment effectiveness. International Journal of Production Research. 2011;**49**:37-41. DOI: 10.1080/00207543.2010.519358

[2] Suryaprakash M, Prabha MG, Yuvaraja M, Revanth RVR. Improvement of overall equipment effectiveness of machining Centre using tpm. Materials Today Proceedings. 2020; **46**:9348-9353. DOI: 10.1016/j.matpr.2020.02.820

[3] Brah SA, Chong W. Relationship between total productive maintenance and performance. International Journal of Production Research. 2004; **42**:37-41. DOI: 10.1080/00207540410001661418

[4] David JMM, Rodriguez RS. Total productive maintenance (TPM) as a tool for improving productivity : A case study of application in the bottleneck of an auto-parts machining line. International Journal of Advanced Manufacturing Technology. 2017;**92**:1013-1026. DOI: 10.1007/s00170-017-0052-4

[5] Ribeiro IM, Godina R, Pimentel C, Silva FJG, Matias JCO. Implementing TPM supported by 5S to improve the availability of an automotive production line. Procedia Manufacturing. 2020; **38**(2019):1574-1581. DOI: 10.1016/j.promfg.2020.01.128

[6] Congress W, Wce VIII, Shah B. Total productive maintenance: A study of Malaysian automotive SMEs. Proceedings of the World Congress on Engineering. 2012;**III**:2-6

[7] Singh R, Gohil AM, Shah DB, Desai S. Total productive maintenance (TPM) implementation in a machine shop: Acase study. Procedia Engineering. 2013; **51**:592-599. DOI: 10.1016/j.proeng.2013.01.084

[8] Munir MA, Zaheer MA, Haider M, Rafique MZ, Rasool MA, Amjad MS. Problems and barriers affecting total productive maintenance implementation. Engineering, Technology & Applied Science Research. 2019;**9**(5):4818-4823

[9] Mishra RP, Gupta G, Sharma A. Development of a model for total productive maintenance barriers to enhance the life cycle of productive equipment. Procedia CIRP. 2021;**98** (March):241-246. DOI: 10.1016/j.procir.2021.01.037

[10] Wakjira MW, Singh AP. Total productive maintenance: A case study in manufacturing industry. Global Journal of Research in Engineering Industrial Engineering. 2012;**12**(1):25-32

[11] Nallusamy S, Kumar V, Yadav V, Prasad UK, Suman SK. Implementation of total productive maintenance to enhance the overall equipment effectiveness. International Journal of Mechanical and Production Engineering Research and Development (IJMPERD). 2018;**8**(1):1027-1038

[12] Graisa M, Al-habaibeh A. An investigation into current production challenges facing the Libyan cement industry and the need for innovative total productive maintenance (TPM) strategy. Journal of Manufacturing Technology Management. 2011;**22**(4): 541-558. DOI: 10.1108/17410381111126445

[13] Ngoy KR, Israel K. The strategy of successful total productive maintenance (TPM): Implementation and benefits of TPM (literature review). International

Journal of Innovative Research in Engineering & Multidisciplinary Physical Sciences. 2021;**9**(6):43-52

[14] Jain A, Bhatti R. Total productive maintenance (TPM) implementation practice a literature review and directions. International Journal of Lean Six Sigma. 2014;5(3):293-323

[15] Mckone KE, Schroeder RG, Cua KO. The impact of total productive maintenance practices on manufacturing performance. Journal of Operations Management. 2001;**19**:39-58

[16] Alhouli Y, Elhag T, Alardhi M, Alazemi J. Development of conceptual framework for ship maintenance performance measurements. Journal of Mechanical Engineering and Automation. 2017;7(3):63-71. DOI: 10.5923/j.jmea.20170703.01

[17] Muchiri P, Pintelon L, Gelders L, Martin H. Development of maintenance function performance measurement framework and indicators. International Journal of Production Economics. 2011; **131**(1):295-302. DOI: 10.1016/j.ijpe.2010.04.039

[18] Yulian E, Adesta T, Prabowo HA. Total productive maintenance (TPM) implementation based on lean tools in Indonesian manufacturing industries. International Journal of Engineering & Technology. 2018;7:156-159. DOI: 10.14419/ijet.v7i3.7.16261

[19] Thomas AJ, Jones GR, Vidales P. An integrated approach to TPM and six sigma development in the castings industry. In: Intelligent Production Machines and Systems. Elsevier; 2006. pp. 620-625

[20] Sharma RK, Sharma RG. Integrating six sigma culture and TPM framework to improve manufacturing performance in SMEs. Quality and Reliability Engineering International. 2014;**30**: 745-765. DOI: 10.1002/qre.1525

[21] Lai CTA, Jiang W, Jackson PR. Internet of things enabling condition-based maintenance in elevators service. Journal of Quality in Maintenance Engineering. 2018;**25**:563-588. DOI: 10.1108/JQME-06-2018-0049

[22] Almeanazel OTR. Total productive maintenance review and overall equipment effectiveness measurement. Jordan Journal of Mechanical and Industrial Engineering. 2010;**4**(4): 517-522

[23] Alsyouf I. The role of maintenance in improving companies' productivity and profitability. International Journal of Production Economics. 2007;**105**:70-78. DOI: 10.1016/j.ijpe.2004.06.057

[24] Rotab Khan MR, Darrab IA. Development of analytical relation between maintenance, quality and productivity. Journal of Quality in Maintenance Engineering. 2010;**16**: 341-353. DOI: 10.1108/13552511011084508

Chapter 4

Assigning Multi-Attribute Targets under Uncertainty with Multiple Suppliers and Stakeholders

Robert F. Bordley

Abstract

A manager must create deliverables satisfying multiple stakeholders. Different stakeholders use the deliverables to satisfy multiple different (and potentially conflicting) objectives. For each criterion, the manager assigns different targets to different suppliers. If suppliers meet their targets, the deliverable score is the sum of the targets. The manager will only be successful if all suppliers meet their targets and if the deliverable scores are acceptable to stakeholders. This chapter solves for the target maximizing the manager's probability of success both in simple cases and in the general case of multiple suppliers, multiple attributes, and multiple stakeholders. The targets introduce a supplier margin of safety (or buffer) to allow for uncertainty in that supplier's capability. The targets also introduce a stakeholder margin of safety to allow for uncertainty in what is acceptable to each stakeholder. The assigned targets reflect trade-offs between the margins of safety for each supplier and stakeholder. For simplicity, all uncertainties are assumed Gaussian, and deliverable scores are additive in the targets. Generalizing both assumptions is straightforward.

Keywords: targets, safety margin, stakeholder, supplier, uncertain needs, uncertain quality

1. Introduction

A project is a temporary endeavor aimed at creating some deliverable subject to cost and deadline constraints. Stakeholders use these deliverables to achieve their own objectives under various scenarios (e.g., a vehicle enables its driver to commute to work and travel medium distances during normal or increment weather.)

The stakeholders and project manager agree on the multi-attribute targets (on multiple attributes) against which deliverable performance will be compared. Complex projects must be decomposed into work packages by dividing each project target into specialized targets for each work package. For example, designing a vehicle to meet some weight target is decomposed into designing tires, seats, engines, exterior panels, etc. to meet weight targets for tires, seats, engines, etc. In this case, the sum of the work package weight targets equals the overall target. Sometimes several tiers of decomposition are required before each work package is sufficiently granular to be assigned to a small specialized supplier.

There are many technical, raw material, personnel, and other issues, which could prevent the supplier from meeting the target. These issues create uncertainties about whether suppliers can meet their targets. To reduce the risk of supplier failure, the manager tracks the progress of each supplier and reallocates resources when necessary. Despite these efforts, however, there will still be some risk of suppliers not meeting their targets.

The manager is also uncertain about whether the agreed-upon overall target accurately reflects what stakeholders need to achieve their objectives. Use cases, sequence diagrams, activity diagrams, and other tools of model-based systems engineering [1] help stakeholders identify all the specific functions the product must perform to help them achieve their objectives. But when projects are lengthy, stakeholder preferences can change because of changes in family size, competitor offerings, environmental threats and opportunities, etc. This can be especially difficult for consumers. To better estimate consumer needs, conjoint measurements combine stakeholder feedback in hypothetical settings with parametric models of stakeholder behavior. There are many other approaches to requirements elicitation [2–5]. But even with these models, there is still uncertainty about what stakeholders will need.

In some cases, stakeholders will request changes in the targets for evaluating the deliverable before the deliverable is finished. This "requirement volatility" is a leading cause of project failure, especially with software projects [6–10]. But if needs change after the project is finished (and the suppliers are compensated), the stakeholder may be dissatisfied and seek an early replacement for the deliverable.

This chapter sets supplier targets given uncertainty about what suppliers can produce and what stakeholders need. When all uncertainties are Gaussian, the solutions will depend on a "safety margin" for suppliers and a "safety margin" for stakeholders.

2. One design team and one stakeholder

2.1 Mass reduction

To reduce the costs of fueling vehicles, customers prefer more fuel-efficient, lower-weight vehicles. While the project team can reduce the weight of a vehicle, there is a limit to how much they can reduce the weight without reducing performance. Suppose this lower limit is a known value, m. Suppose that the greatest weight customers will accept is known to equal M. Then if $m < M$, the design team can meet customer requirements.

But customer demand for lower-weight vehicles is heavily affected by fuel costs. Fuel costs can change dramatically as the supply of fuel changes. Over the past century, discoveries of new oil fields and the invention of fracking have expanded the availability of fuel. In addition, the development of fuel-efficient engines (including hybrid and electric vehicles) also reduces the number of fuel required to reach a destination. These factors might cause customers to accept vehicle weights higher than M. But oil production often varies, sometimes in response to dwindling oil reserves and sometimes to optimize revenues. Taxes on fuel also increase the cost of using fuel to travel to a destination. These factors could make customers more likely to insist on weights less than M. Both of these factors create uncertainty about how much the maximum weight customers will accept differs from M. Suppose this uncertainty is described by a normal distribution with mean M and standard deviation S. To elicit S, it is common [9] to ask consumers to specify

- A pessimistic but realistic ("rainy day") estimate of maximum acceptable weight when fuel costs are high. (The chances of customers requiring an even lower weight are negligible.)

- An optimistic but realistic ("sunny day") estimate of maximum acceptable weight when fuel costs are low. (The chances of customers being will to accept an even higher weight are negligible.)

This definition presumes a quantitative definition of "negligible chance." Hypothesis testing in the social sciences typically defines a confidence/credibility interval to be approximately four standard deviations in length[1]. As a result, the standard deviation, S, is one-quarter of the absolute value of the difference (the range) between pessimistic and optimistic. The expected value, M, is the average of the bounds.

Let $\Phi(z)$ be the probability a normally distributed value with zero mean and unit standard deviation is less than z. Since the customer is more likely to accept lower-mass targets t, define:

Stakeholder margin: The difference, $M - t$, between the best estimate of the highest acceptable mass and the target.

The probability of a target t, if achieved, being acceptable to the customer is $\Phi\left(\frac{M-t}{S}\right)$.

Suppose the manager recognizes the variation in customer needs but ignores the variation in what their team can achieve. Then since, they believe their team can design a vehicle with mass m, they set $t = m$ and erroneously estimate the probability of success as the probability of $t = m$ satisfying the customer. Then $z = \frac{M-m}{S}$ and $\Phi(z) > \frac{1}{2}$ since $M > m$.

In reality, uncertainties in material costs, technology readiness, available personnel, etc. create uncertainty about how much the total project weight can be reduced. To quantify this uncertainty, the supplier is asked to specify:

- A pessimistic ("rainy day") estimate of the minimum weight that can be supplied if everything goes wrong. (The chances of being unable to provide a weight below this bound are negligible.)

- An optimistic ("sunny day") estimate of the minimum weight that can be supplied if everything goes right. (The chances of being able to provide a weight below this bound are negligible.)

If the uncertainty is normally distributed, then one-quarter of the absolute difference between these bounds (the range) is an estimate of s, the standard deviation of the uncertainty. The average of the bounds is an estimate of the expected value (or m). Since the team is more likely to achieve higher targets, define:

---

[1] The actual range of the traditional 95% confidence/credibility interval is approximately 3.92 standard deviations. For simplicity, this chapter uses four standard deviations which corresponds to a 95.5% interval.

	Supplier	Customer
Lower bound	1363	400
Upper bound	1803	2200
Mean	1583	1300
Sigma	110	450

Table 1.
Data: one supplier, one user.

	Supplier	User
Target	1677	1677
Margin	94	−377
Z-Score	.85	−0.83
Probability	0.80	0.20
with an overall probability of success of 16%. *Thus, failure is likely.*		

Table 2.
Solution: one supplier, one user.

Supplier margin: The difference between the target and the best estimate of the least possible mass, $t - m$.

Then the probability of the supplier being able to achieve the target is $\Phi\left(\frac{t-m}{s}\right)$. Increasing t increases supplier margin but lowers stakeholder margin. To balance these margins, the manager chooses the target to maximize the joint probability of the target being both feasible and sufficient. If the uncertainties in the customer environment are independent of the uncertainties in the supplier environment, this probability is

$$\Phi\left(\frac{t-m}{s}\right)\Phi\left(\frac{M-t}{S}\right) \tag{1}$$

2.2 Numerical example

Table 1 presents the supplier and customer upper and lower bounds along with computations of their means and standard deviations (denoted by σ).

Table 2 presents the results of using excel solver to identify optimal targets with the targets, margins, z-scores, and probability of success.

3. Multiple suppliers

3.1 General solution

Having a single large supplier work, a single project is often inefficient. For example, a meeting with twenty people often devolves into a meeting between smaller subgroups, with other people uninvolved. As a result, best practices in project management and systems engineering decompose the work into smaller work packages. This decomposition continues into every more granular work packages until each of the work packages is sufficiently specific to be addressed by a small specialized supplier.

Suppose the automobile weight reduction problem is decomposed into weight reduction targets for N different automotive components. The smallest feasible weight for work package j is normally distributed with mean m_j and standard deviation s_j for $j = 1, \dots N$. If weight targets $t_1, \dots t_N$ are assigned to each of the component suppliers, the probability all suppliers can meet their targets and satisfy the customer is

$$\prod_{k=1}^{N} \Phi\left(\frac{t_k - \mu_k}{\sigma_k}\right) \Phi\left(\frac{M - \sum t_k}{\sqrt{s^2 + S^2}}\right) \tag{2}$$

3.2 Numerical example

Suppose the vehicle is broken up into three components: Powertrain, Chassis, and Body. Suppose that all component suppliers might have spent 10 hours in a large meeting — with 2 hours focused on discussing the work of each small team, 1 hour focused on the three interactions between each pair of teams, and 1 hour focused on the joint interactions between all suppliers. Then distribution of the work to each separate supplier reduces the time each supplier spends in irrelevant meetings by 4 hours.

If the average mass reduction from reducing meeting time by 4 hours is $\frac{10}{6}$, then the minimum total feasible weight from three more specialized suppliers is 950. The proportion of average weight reduction from each supplier remains the same as in **Table 1** (with standard deviations unchanged.) Upper and lower bounds can then be computed from summing the mean and adding (or subtracting) two standard deviations.

Based on the first numerical example, **Table 3** specifies lower and upper four-sigma bounds on the minimum mass feasible and the maximum mass acceptable for each of the three components and the user. This is used to compute means, standard deviations, and, for an initial set of targets, z-scores, margins, and success probabilities (feasibility probabilities for each design team, acceptability probabilities for the user, and the overall probability of feasibility and acceptability.)

There are uncertainties in what is physically feasible and what the user considers acceptable.

If both uncertainties are ignored, then the manager can set any targets satisfying $t_P > m_P, t_C > m_C, t_B > m_B, M \geq t_P + t_C + t_B$.

If only uncertainty about feasibility is ignored, then the manager will minimize weight by assigning $t_P = m_P, t_C = m_C, t_B = m_B$. **Table 4** specifies these initial targets, margins, z-scores, and probabilities:

	Powertrain	Chassis	Body	User
Lower bound	175	50	400	400
Upper bound	3254	150	800	2200
Mean	250	100	600	1300
Sigma	37.5	25	100	450

Table 3.
Data: multiple suppliers, single user.

	Powertrain	Chassis	Body	User
Target	250	100	600	950
Margin	0	0	0	250
Z-score	0	0	0	0.7777
Probability	0.5	0.5	0.5	0.78

Table 4.
When supplier uncertainty is ignored.

	Powertrain	Chassis	Body	User
Target	340	164	796	1300
Margin	90	64	196	0
Z-score	2.4	2.6	1.96	0
Probability	0.99	0.99	0.97	0.5
with an overall success probability of 48%.				

Table 5.
When user uncertainty is ignored.

	Powertrain	Chassis	Body	User
Target	325	154	746	1225
Margin	75	55	146	75
Z-Score	2.0	2.2	1.5	2.0
Probability	0.98	0.99	0.93	0.90
with an overall success probability of 87%.				

Table 6.
Solution: multiple suppliers, single user when no uncertainties are ignored.

The suppliers get zero margin and the customer gets a margin of 250. Thus, the feasibility probabilities for the suppliers are $\frac{1}{2}$, while the acceptability probability for the customer is 78%. Multiplying these probabilities gives an overall success probability of 9.75%.

Alternatively, suppose the feasibility uncertainty is considered, while uncertainty about customer acceptance is ignored. In this case, the project manager requires that $t_P + t_C + t_B = M(1300)$ while setting t_P, t_C, t_B to maximize the probability of feasibility. The targets, margins, z-scores, and probabilities in **Table 5**.

Setting targets to recognize uncertainty in both feasibility and customer acceptance requires relaxing the constraint that the sum of the supplier targets equals the average customer need, M. **Table 6** presents the solution without this constraint.

4. Multiple attributes and stakeholders

4.1 Multiple attributes

Typically targets are set on multiple attributes, e.g., weight, safety, comfort, convenience, speed, etc. If uncertainty on each attribute k is measured by a Gaussian distribution and all attributes are uncorrelated, then targets will be set to maximize

$$\prod_{k=1}^{N} \Phi\left(\frac{t_k - m_k}{s_k}\right)\Phi\left(\frac{M_k - t_k}{S_k}\right). \tag{3}$$

But physical improvements on one attribute often affect the ability to physically improved on another attribute. Suppose these physical interactions are described by a multivariate normal with mean vector $\vec{m}$ and variance-covariance $\mathbf{s}$.

Likewise, there are interactions between improvements in one attribute and preferences for other attributes. For example, improving a vehicle's crashworthiness and its ability to avoid crashes both increase the safety of passengers in the vehicle. Hence, dramatic improvements on one attribute reduce the need to make improvements on other attributes, that is. the attributes are partially substitutable. In contrast, improving a vehicle's reliability increases the amount of time a driver can use the vehicle which increases the need for performance improvements. These attributes are complementary. Suppose these correlations between preference attributes are described by a multivariate normal distribution with mean $\vec{M}$ and variance-covariance $\hat{\mathbf{S}}$. Then if there is still no correlation between customer preferences and supplier uncertainties, targets must be set to maximize

$$N\left(\vec{t} - m, \mathbf{S}\right)N\left(\vec{m} - t, \mathbf{s}\right). \tag{4}$$

If the standard deviation on a requirement is small, then the normal distribution will attach high weight to improving performance to meet the target. Thus, the normal distribution also allows for the prioritization of different attribute targets.

Some suppliers, for example, powertrain will have a significant impact on meeting vehicle emissions and fuel-efficiency targets but not on comfort and convenience. Other suppliers, for example, body will have less impact on emissions and fuel efficiency but greater impact on comfort and convenience. Software, while having a great impact on many performance targets, typically has a negligible impact on vehicle weight. This formulation allows target allocation to reflect these differences among different work teams.

4.2 Multiple stakeholders

In many cases, stakeholders have very different, or even conflicting, objectives. Stakeholders concerned with vehicle crashworthiness might prefer higher-mass vehicles to lower-mass vehicles. Suppose their minimum acceptable mass has mean M_2 and standard deviation S_2 (with the original stakeholders having mean M_1 and standard deviation S_1.) If we also use the numbers 1, 2, 3 to represent the powertrain, chassis, and body group, then the manager must set targets to maximize

$$\left[\prod_{i=1}^{3}\Phi\left(\frac{t_i - m_i}{s_i}\right)\right]\Phi\left(\frac{M_1 - \sum_i t_i}{S_1}\right)\Phi\left(\frac{\sum_i t_i - M_1}{S_2}\right). \quad (5)$$

This distinction between stakeholders becomes especially important in setting multi-attribute targets.

Different stakeholders are often only interested in different targets. Government regulators are typically interested in vehicle emissions but not comfort, convenience, and style. Buyers are interested in comfort, convenience, and style but not emissions. There are some attributes, for example, vehicle safety and fuel efficiency where both are interested.

4.3 Example

Suppose there are two engineering groups (Powertrain, Body), two user groups (Customer, Govt), and two attributes (Defect rate, Weight). The customer demands average defect rates lower than the government, while the government demands average weights lower than the customer. The powertrain team provides a lower average defect rate, while the body group provides a lower average weight. For each attribute, **Table 7** lists the capability and needs of engineering and users.

Table 8 presents the results of using an excel solver to identify optimal targets with the targets, margins, z-scores, and probability of success.

	Powertrain	Body	Customer	Govt
Lower bound: Defects	2	10	16	20
Upper bound: Defects	8	26	36	50
Mean: Defects	5	17	27	35
Sigma: Defects	1.5	4	5	7.5
Lower bound: Weight	175	50	350	250
Upper bound: Weight	325	150	750	550
Mean: Weight	250	100	550	400
Sigma: Weight	37.5	25	100	75

Table 7.
Data: multiplier suppliers, attributes, users.

	Powertrain	Body	Customer	Govt
Target: Defects	6.8	18.5	25.3	25.3
Margin: Defects	1.8	1.5	1.7	9.7
Z-score: Defects	1.2	0.4	0.3	1.3
Probability: Defects	0.90	0.64	0.63	0.90
Target: Weight	293	136	429	429
Margin: Weight	43	36	121	71

DOI: http://dx.doi.org/10.5772/intechopen.114376

	Powertrain	Body	Customer	Govt
Z-score: Weight	1.1	1.4	1.2	0.95
Probability: Weight	0.87	0.92	0.89	0.93

Table 8.
Solutions: multiple suppliers, attributes, users.

For an overall 33 and 59% probability of meeting the defect and weight targets, respectively. Because devoting time to defect reduction might divert effort from weight reduction, there might be a negative correlation between each engineer team's capability on each attribute. Since customers might feel willing to accept higher weight in return for few defects, there might also be negative correlation between satisfaction with the attributes. Extensions of this approach could address these issues.

5. Conclusions

This chapter illustrated how targets could be set to balance feasibility and sufficiency for multiple suppliers and multiple stakeholders. Initial examples focused on the case of one supplier and one stakeholder and showed how ignoring either supplier uncertainty or stakeholder uncertainty reduces the probability of success. This illustrated the importance of both stakeholder and supplier margins.

However, even fully considering both uncertainties still did not lead to a high probability of success — because of the inefficiency of one team addressing a complex problem. This motivated the decomposition of a single supplier into multiple more specialized suppliers each focused on a different work package.

This approach was extended to multi-attribute targets with complementary and substitutable attributes. Finally, it was extended to multiple stakeholders, which allowed consideration of differences between stakeholder preferences for different attributes. Thus, this chapter presents a tractable approach of addressing complex interactions between design sub-teams, attribute targets, and the preferences of different stakeholders in the presence of uncertainty.

Behavioral research finds that individuals naturally evaluate outcomes against reference points and that how a product is described can influence individual reference points. Hence, this approach highlights the role of advertising and social media in shaping what individuals find acceptable. This joint use of product design and marketing can then increase the probability of success.

While this paper assumed Gaussian distributions, extensions to more realistic distributions are straightforward. It is also straightforward to generalize this chapter's assumption of an additive relationship between supplier targets and the overall target.

Author details

Robert F. Bordley
Integrative Sciences Division, College of Engineering, University of Michigan, Ann Arbor, United States

*Address all correspondence to: rbordley@umich.edu

References

[1] Micouin P. Model-Based Systems Engineering: Fundamentals and Methods. Hoboken, New Jersey: John Wiley & Sons; 2014. ISBN: 978-1-848-21469-9

[2] Christel M, Kang K. Issues in Requirements Elicitation. Technical Report. Pittsburgh, Pennsylvania: Technical Report, Carnegie-Mellon Software Institute; 1992

[3] Davis A, Dieste O, Hickey A, Juristo N, Moreno AM. Effectiveness of requirements elicitation techniques: Empirical results derived from a systematic review. In: IEEE International Conference on Requirements Engineering, Minneapolis, Minnesota. Vol. 11. 2006. pp. 176-185

[4] Dias L, Morton A, Quigley J. Elicitation: The Science and Art of Structuring Judgement. New York City, New York: Springer; 2018

[5] Ferrari A, Spoletini P, Debnath S. How do requirements evolve during elicitation? An empirical study combining interviews and app store analysis. In: Requirements Engineering. Vol. 27. New York City, New York: Springer; 2022. pp. 489-519

[6] Abd K, Abd M, Kholeif S. Identify and manage software requirements volatility. International Journal of Advanced Computer Science and Applications. 2016;7(5):64-71

[7] Bano M, Imtiaz S, Ikram N, Usman M, Niazi M. Systematic literature review for causes of requirements change. Technical Report: TR/2010–01, School of Computing and Mathematics, Keele University, Keele, UK. 2010. ISSN: 1353:7776

[8] Bowen P, Heales J, Vongphakdi M. Reliability factors in business software: Volatility, requirements and end-users. Information Systems Journal. 2002;**12**(3):186-213

[9] Ferreira S, Collofello J, Schunk D, Mackulak G. Understanding the effects of requirements volatility in software engineering. Journal of Systems and Software. 2009;**82**(10):1568-1577

[10] Kulk G, Verhoef C. Quantifying requirements volatility effects. Science of Computer Programming. 2008;**72**(3):136-175

Section 2

Applications

Chapter 5

Advanced System Diagnostics Tools: Innovations and Applications

Adeel H. Suhail, Fiseha M. Guangul and Abdul Nazeer

Abstract

This chapter explores the evolution and significance of advanced diagnostics tools in complex system management, emphasizing the shift toward integrated, intelligent, and predictive diagnostics. It covers the application of artificial intelligence (AI) and machine learning (ML) for predictive maintenance, real-time monitoring, and the integration of sensor technologies with the Internet of Things (IoT). The text examines the use of diagnostic tools in various fields, including mechanical, electrical, software, and network systems, with specific attention to industry applications in automotive, aerospace, and healthcare sectors. These examples illustrate how diagnostics are transforming these industries by enhancing efficiency, safety, and reliability. The chapter also discusses the role of diagnostics in system design, the importance of built-in diagnostics, and the challenges of managing and interpreting the large volumes of data these tools generate. It highlights innovative data analysis and visualization techniques and looks ahead to future trends in diagnostics, such as the potential impact of quantum computing and concerns over cybersecurity and data privacy. Overall, the chapter provides a comprehensive overview of the current state and future directions of advanced system diagnostics.

Keywords: advanced diagnostic tools, real-time monitoring, data analytics in diagnostics, integration of IoT in systems, artificial intelligence in diagnostics

1. Introduction

This chapter provides an in-depth exploration of the latest diagnostic technologies, including the integration of artificial intelligence (AI) and machine learning (ML), real-time monitoring, predictive maintenance, and the burgeoning role of Internet of Things (IoT) in system health assessment. We examine a range of tools tailored for diverse systems, from mechanical and electrical to software and network diagnostics, highlighting their critical role in various industries such as automotive, aerospace, and healthcare [1].

The chapter not only focuses on the current state-of-the-art tools but also discusses the integration of these diagnostics into system design, emphasizing the importance of built-in diagnostics and feedback loops for continuous monitoring and improvement. Additionally, we tackle the challenges of handling and interpreting the vast amounts of data generated by these tools, showcasing advanced techniques in data analysis and visualization that aid in decision-making.

Looking toward the future, the chapter anticipates emerging trends and challenges, including the potential impact of quantum computing and the growing concerns around cybersecurity and data privacy in system diagnostics [2]. This comprehensive overview is designed to equip readers with a thorough understanding of the current landscape and future directions in advanced system diagnostics, emphasizing their pivotal role in enhancing the efficiency, safety, and reliability of complex systems.

1.1 Overview of system diagnostics

System diagnostics plays an essential role in the maintenance and optimization of complex engineering systems. It encompasses the processes, methods, and technologies used to identify, assess, and rectify malfunctions within these systems, ensuring they operate at optimal efficiency and reliability. The scope of system diagnostics spans across various fields, including automotive, aerospace, industrial automation, and healthcare, highlighting its significance in modern engineering practices. This foundational aspect of engineering not only aids in preventive maintenance but also supports the development of more resilient and efficient systems. By accurately diagnosing issues, engineers can prevent potential failures, minimize downtime, and extend the lifecycle of critical components, thereby safeguarding both the functionality and integrity of the system as a whole.

The primary objective of system diagnostics is to facilitate preventive maintenance, ensuring that systems operate without unforeseen failures that could lead to costly downtime or hazardous situations. Through the early detection and correction of faults, diagnostics serve as a proactive measure rather than a reactive one, significantly enhancing system reliability and safety. Moreover, diagnostics play a pivotal role in performance optimization, allowing for the fine-tuning of systems to achieve superior operational efficiency. By continuously monitoring system parameters and performance indicators, diagnostics tools enable engineers to make informed decisions, leading to improved system performance and longevity. This preventive approach not only reduces maintenance costs but also contributes to the sustainability of engineering systems by optimizing resource use and minimizing waste.

Predictive maintenance, powered by ML algorithms, enables proactive identification of potential system failures before they occur. This approach not only enhances system reliability but also optimizes maintenance schedules, reducing downtime and operational costs. The integration of sensor data with ML models facilitates the early detection of anomalies, allowing for timely interventions [3, 4].

1.2 Understanding the role of diagnostics in systems engineering

In the realm of systems engineering, diagnostics is not merely an add-on but a fundamental component that influences every stage of a system's lifecycle. Its integration into system design from the outset ensures that maintenance and fault rectification are considered early, thereby embedding reliability and ease of repair into the system architecture [5]. Diagnostics enhances system reliability and safety by providing the necessary tools and information to detect and rectify faults promptly, preventing minor issues from escalating into major system failures [6]. This proactive approach to maintenance and repair not only extends the operational life of the system but also ensures that it meets the stringent safety standards required in many engineering fields, thereby protecting both assets and lives.

Diagnostics significantly impact the development, operation, and maintenance phases of system lifecycle management. During the development phase, diagnostics tools and techniques are used to validate design assumptions, verify system functionality, and ensure that the system meets all specified requirements [7]. Once the system is operational, diagnostics facilitate ongoing maintenance and troubleshooting, enabling the timely detection and resolution of issues that may arise. This ongoing engagement with the system's operational health not only ensures its reliability and efficiency but also provides valuable data that can inform future design improvements and innovations [8]. By fostering a continuous feedback loop between operation, maintenance, and design, diagnostics contribute to the iterative improvement of systems engineering practices and outcomes.

The interdisciplinary approach in system diagnostics underscores the integration and collaboration between various engineering disciplines to tackle complex diagnostic challenges [9]. This collaborative effort is vital as the complexity of modern systems often transcends the boundaries of a single engineering field, encompassing mechanical, electrical, software, and chemical components. For instance, the diagnostic of an autonomous vehicle involves mechanical engineers for the vehicle's physical systems, electrical engineers for the sensor and actuator systems, and software engineers for the onboard computing systems. By fostering a collaborative environment, an interdisciplinary approach enables the synthesis of diverse expertise, leading to more comprehensive and effective diagnostic strategies. This not only enhances the reliability and performance of systems but also drives innovation in diagnostic tools and methodologies, benefiting from the unique perspectives and knowledge bases of different disciplines.

The synergy between system diagnostics and information technology (IT) is a hallmark of the interdisciplinary approach, highlighting how IT tools and methodologies can significantly enhance diagnostic capabilities. Advances in data analytics, cloud computing, and ML, for instance, have opened new avenues for diagnosing complex systems [10]. By leveraging big data analytics, engineers can process and analyze vast amounts of diagnostic data in real time, identifying patterns and anomalies that may indicate potential issues. Cloud computing facilitates the remote monitoring and diagnostics of systems across different locations, enhancing the efficiency of maintenance operations. Moreover, ML algorithms can predict potential failures and suggest preventive measures, improving the decision-making process in maintenance management. This convergence of engineering and IT not only amplifies the effectiveness of diagnostics but also underscores the need for professionals who are adept in both fields, highlighting the importance of interdisciplinary education and training in preparing the next generation of engineers.

1.3 Evolution of diagnostic tools and techniques

The evolution of diagnostic tools and techniques is a testament to the advancements in engineering and technology over the years. Initially, diagnostics relied heavily on manual inspection and simple testing equipment, which were time-consuming and often inaccurate. The development of more sophisticated tools, such as oscilloscopes and multimeters, marked a significant step forward, enabling more precise and efficient fault detection. As technology progressed, so did the complexity of systems and the need for more advanced diagnostics solutions. This historical journey from rudimentary tools to complex diagnostic systems reflects the growing demands of modern engineering and the continuous drive for innovation in the field.

The technological revolution, characterized by the rise of digital technologies, computational power, and connectivity, has dramatically transformed the landscape of system diagnostics. The introduction of ML and AI has ushered in a new era of predictive diagnostics, enabling the anticipation of failures before they occur. Similarly, the IoT has revolutionized remote monitoring, allowing for real-time data collection and analysis from systems across the globe. These advancements have not only increased the accuracy and efficiency of diagnostics but also made them more accessible and cost-effective. As we look to the future, emerging trends such as quantum computing hold the promise of further enhancing diagnostic capabilities, potentially revolutionizing how we approach maintenance and fault detection in complex systems [11, 12].

2. Cutting-edge diagnostic technologies

2.1 AI and ML in diagnostics

Artificial intelligence (AI) and machine learning (ML) have become cornerstones of modern diagnostic technologies, transforming the landscape of system maintenance and repair. These technologies enable the analysis of vast datasets, identifying patterns and anomalies that elude traditional diagnostic methods. AI algorithms can predict failures before they occur, allowing for interventions that can prevent downtime and extend equipment life. ML models, trained on historical performance and maintenance data, continuously improve, becoming more accurate in their predictions over time. This adaptive learning capability is particularly beneficial in complex systems where dynamic changes and unpredictable variables can complicate diagnostics. The application of AI and ML in diagnostics represents a shift toward more intelligent, efficient, and predictive maintenance strategies, significantly reducing the reliance on reactive maintenance practices [13].

2.2 Real-time monitoring and predictive maintenance

Real-time monitoring and predictive maintenance technologies represent a significant advancement in diagnostics, moving beyond periodic checks to continuous system surveillance. This approach utilizes a network of sensors and IoT devices to collect data on system performance in real time. Advanced analytics tools then process this data, identifying trends and predicting potential failures before they disrupt operations. This predictive maintenance strategy not only optimizes the maintenance schedule based on actual system needs but also significantly reduces unplanned downtime and maintenance costs. Moreover, real-time monitoring enables the immediate detection of anomalies, facilitating swift interventions that can prevent minor issues from escalating into major system failures. The integration of real-time data collection with predictive analytics embodies the proactive spirit of modern diagnostics, ensuring system reliability and performance are maintained at optimal levels.

2.3 Sensor technologies and IoT integration

Sensor technologies and the integration of the IoT are pivotal to the advancement of cutting-edge diagnostic solutions. Sensors, capable of detecting a wide range of physical, chemical, and operational parameters, serve as the primary data sources for diagnostic systems. When integrated with IoT technology, these sensors become part of a connected ecosystem that can communicate data in real time to centralized analytics platforms. This integration facilitates the continuous monitoring of systems across multiple locations, enabling a holistic approach to diagnostics and maintenance. IoT connectivity also supports the remote diagnosis of issues, allowing maintenance teams to assess system health and perform troubleshooting without the need for on-site visits. The widespread adoption of sensor technologies and IoT integration is driving the development of smarter, more interconnected diagnostic systems that can anticipate and respond to system needs with unprecedented precision and efficiency [14].

3. Diagnostic tools for complex systems

3.1 Tools for mechanical and electrical systems

Diagnostic tools designed for mechanical and electrical systems have evolved significantly, incorporating sophisticated technologies to ensure the optimal performance and reliability of such systems. For mechanical systems, vibration analysis tools, thermal imaging cameras, and ultrasonic detectors are commonly used to predict and prevent failures by detecting anomalies such as misalignments, imbalances, and bearing failures. Electrical systems benefit from advanced multimeters, oscilloscopes, and power quality analyzers that can diagnose issues ranging from simple electrical faults to complex power distribution problems. The integration of these tools with advanced data analytics further enhances their diagnostic capabilities, allowing for the early detection of potential failures and the implementation of corrective actions before costly downtimes occur. This integrated approach not only increases the longevity of mechanical and electrical components but also ensures that systems operate within their optimal performance parameters, contributing to overall system efficiency and safety [15].

3.2 Software diagnostics: debugging and performance monitoring

In the realm of software diagnostics, debugging tools and performance monitoring solutions play critical roles in ensuring the reliability and efficiency of software systems. Debugging tools, such as integrated development environments (IDEs) and specialized debugging software, allow developers to systematically identify and correct code errors, ensuring that software functions as intended. Performance monitoring tools, on the other hand, provide real-time insights into software operation, tracking metrics such as response times, system resource usage, and throughput rates. These tools are essential for identifying performance bottlenecks and optimizing software for better efficiency and scalability. The use of AI and ML in software diagnostics is also on the rise, offering predictive insights into potential issues and automating aspects of the debugging and optimization processes. As software systems become increasingly complex, the importance of advanced software diagnostics tools in maintaining system integrity and performance cannot be overstated [16].

3.3 Network diagnostics: ensuring system connectivity and security

Network diagnostics tools are essential for maintaining the connectivity and security of digital communication networks. These tools range from simple ping and traceroute commands to more sophisticated network analyzers and monitoring software that can map network topology, monitor traffic in real time, and identify potential security breaches. With the growing threat of cyberattacks, security diagnostic tools such as intrusion detection systems (IDS) and network firewalls have become indispensable in identifying and mitigating threats. Moreover, the integration of AI and ML in network diagnostics allows for the predictive analysis of network traffic patterns, enabling the proactive identification of anomalies that could indicate security vulnerabilities or performance issues. As networks continue to expand and support an increasing number of connected devices, the role of network diagnostics in ensuring system connectivity and security becomes ever more critical, underscoring the need for continuous advancements in diagnostic technologies and methodologies [17].

4. Case studies in system diagnostics

4.1 Automotive industry: advanced diagnostics in vehicle systems

The automotive industry represents a pioneering sector in the adoption and development of advanced diagnostics, significantly enhancing vehicle systems' safety, efficiency, and reliability. Modern vehicles are equipped with on-board diagnostics (OBD) systems that monitor engine and other vehicle components' performance. These systems can detect malfunctions and alert drivers to potential issues long before a breakdown occurs. Additionally, telematics technologies enable remote diagnostics, allowing manufacturers and service providers to assess vehicle health in real time and even predict future maintenance needs. A notable case is the integration of AI-driven predictive maintenance models that analyze data from various sensors within the vehicle to forecast component failures. Such advancements not only improve vehicle longevity and reduce maintenance costs but also contribute to safer driving experiences by ensuring that potential issues are addressed proactively [18].

4.2 Aerospace diagnostics: ensuring flight safety and efficiency

In the aerospace sector, diagnostics play a crucial role in ensuring the safety and efficiency of air travel. The complexity of aircraft systems necessitates sophisticated diagnostic tools capable of monitoring and analyzing a vast array of parameters in real time. Advanced health monitoring systems, including structural health monitoring (SHM) and engine condition monitoring (ECM), are employed to detect and diagnose potential issues that could compromise flight safety. These systems utilize data analytics, ML, and predictive modeling to anticipate failures before they occur, allowing for timely maintenance and repairs. A key case study involves the use of digital twin technology to create virtual replicas of aircraft components, enabling the simulation and analysis of wear and tear under various conditions. This innovative approach allows engineers to predict failures and optimize maintenance schedules, significantly enhancing aircraft reliability and operational efficiency [19].

4.3 Healthcare systems: diagnostic tools in medical equipment

The healthcare industry has witnessed significant advancements in diagnostic tools, revolutionizing patient care and treatment outcomes. Medical diagnostics now benefit from a wide range of technologies, from imaging equipment such as MRI and

CT scanners to wearable devices that monitor vital signs and detect abnormalities. The integration of AI and ML in diagnostic processes has enabled the development of tools capable of identifying diseases with greater accuracy and speed than ever before. For instance, AI algorithms can analyze medical images to detect cancerous tumors at early stages, significantly improving treatment success rates. Additionally, IoT-enabled wearable devices provide continuous health monitoring, offering real-time data that can be used to diagnose conditions or adjust treatment plans. A notable case study is the development of non-invasive glucose monitoring devices for diabetic patients, allowing for constant blood sugar monitoring without the need for painful and inconvenient blood tests. These advancements in healthcare diagnostics not only improve patient outcomes but also contribute to more personalized and efficient healthcare delivery [20].

5. Integration of diagnostic tools into system design

5.1 Designing systems with built-in diagnostics

Incorporating diagnostics into the initial design phase of systems represents a forward-thinking approach that enhances both functionality and reliability. Designing systems with built-in diagnostics involves embedding sensors, software, and other diagnostic components directly into the system architecture, allowing for continuous monitoring and assessment of system performance. This proactive design strategy enables the early detection of issues, facilitating immediate corrective actions that prevent minor faults from escalating into major failures. It also simplifies maintenance processes, as diagnostic data can guide technicians directly to the source of a problem, reducing repair times and costs. Automotive manufacturers, for instance, integrate extensive diagnostic protocols within vehicle management systems, enabling real-time monitoring of engine performance, emission levels, and electrical systems. This integration not only ensures optimal vehicle performance but also supports compliance with regulatory standards [21].

5.2 The role of feedback loops in system health monitoring

Feedback loops are vital components of integrated diagnostic systems, enabling continuous improvement and adaptation based on real-time data. In systems with built-in diagnostics, feedback loops collect data on operational performance and use it to inform system adjustments, ensuring that systems operate within optimal parameters. This dynamic process of monitoring, analyzing, and adjusting is crucial for maintaining system health and efficiency. For example, in industrial automation, feedback loops can adjust machine operations based on wear and tear indicators, optimizing performance, and extending machinery life. Similarly, in smart buildings, feedback from environmental monitoring systems can automatically adjust heating, ventilation, and air conditioning (HVAC) settings to maintain comfort while optimizing energy use. The effective use of feedback loops in system design not only enhances system reliability and efficiency but also contributes to sustainability by reducing waste and energy consumption [22].

5.3 Challenges and best practices in integration

Integrating diagnostic tools into system design presents several challenges, including technological compatibility, data overload, and ensuring user privacy and data security. Overcoming these challenges requires a comprehensive approach that considers the system's entire lifecycle, from design through to decommissioning. Best practices in integration include adopting modular designs that allow for easy updates and expansions of diagnostic capabilities as new technologies emerge. Ensuring interoperability between different diagnostic tools and system components is also crucial, necessitating standardization efforts and collaboration between manufacturers and software developers [23]. Additionally, managing the vast amounts of data generated by diagnostic tools requires robust data management and analysis frameworks, leveraging cloud computing and advanced analytics to extract actionable insights without overwhelming system operators.

Privacy and security are paramount, particularly in sectors such as healthcare and automotive, where sensitive data is routinely processed. Implementing strong encryption protocols, access controls, and regular security audits can mitigate risks associated with data breaches and unauthorized access. Furthermore, engaging end-users and maintenance personnel in the design process ensures that diagnostic systems are user-friendly and provide meaningful, actionable information, enhancing their effectiveness in real-world applications.

Ultimately, the successful integration of diagnostic tools into system design hinges on a balance between technological innovation and practical usability. By adhering to these best practices, engineers and designers can overcome the inherent challenges, paving the way for the next generation of smart, self-diagnosing systems that promise enhanced reliability, efficiency, and user satisfaction.

6. Data analysis and interpretation in diagnostics

6.1 Handling large volumes of diagnostic data

The advent of sophisticated diagnostic tools across various industries has led to an explosion in the volume of data generated. Handling these large datasets requires advanced data management strategies to ensure that the information is not only stored securely but is also accessible for analysis. Big data technologies and cloud computing platforms play a critical role in managing these data volumes, providing the computational power necessary to process and analyze data in real time. Effective data management also involves implementing data normalization and standardization practices, which facilitate interoperability between different systems and tools. Moreover, employing ML algorithms can help in sifting through vast datasets to identify patterns, trends, and anomalies, transforming raw data into actionable insights. For industries such as healthcare and aerospace, where the stakes are high, the ability to efficiently manage and analyze diagnostic data can significantly impact decision-making processes, system reliability, and safety [24].

6.2 Data visualization techniques for system health

Data visualization is an essential aspect of diagnostics, providing a user-friendly interface for interpreting complex datasets. Effective visualization techniques transform abstract data into intuitive graphical representations, making it easier for engineers, technicians, and decision-makers to understand system health at a glance. Techniques such as dashboards, heat maps, and time-series graphs allow users to monitor system performance in real time, identify trends, and spot anomalies.

For instance, in the automotive industry, visual diagnostics tools can display engine performance metrics, alerting drivers to potential issues before they become serious. In manufacturing, dashboards can visualize production line data, helping managers optimize operations and reduce downtime. The key to effective data visualization lies in simplicity and clarity, ensuring that the most critical information is readily accessible to inform decision-making processes without overwhelming the user with too much detail [25].

6.3 Decision-making based on diagnostic insights

The ultimate goal of diagnostics is to inform and improve decision-making processes, ensuring that systems operate efficiently, safely, and reliably. Diagnostic insights derived from data analysis and visualization tools must be actionable, providing clear guidance on maintenance, repair, and operational adjustments. This decision-making process often involves weighing the costs and benefits of different actions, considering factors such as downtime, repair costs, and potential risks. For example, predictive maintenance strategies informed by diagnostic data can help organizations schedule repairs at the most opportune times, balancing the need for system uptime with maintenance requirements. Additionally, in sectors like healthcare, diagnostic insights can guide treatment decisions, impacting patient outcomes. Decision-making based on diagnostic insights requires a multidisciplinary approach, integrating technical knowledge with operational considerations and strategic planning. As such, fostering collaboration across departments and leveraging expert systems and AI can enhance the decision-making process, ensuring that insights are implemented effectively and contribute to overall system optimization [26].

7. Future trends in system diagnostics

7.1 The potential of quantum computing in diagnostics

Quantum computing holds transformative potential for the field of system diagnostics by offering computational capabilities far beyond those of classical computers. This emerging technology could revolutionize the way diagnostics are conducted, especially in handling complex problems that involve vast amounts of data and require significant computational power. Quantum computing's ability to perform parallel computations and analyze large datasets almost instantaneously makes it particularly suited for predictive diagnostics and the analysis of complex systems. In healthcare, for instance, quantum computing could significantly enhance the accuracy and speed of genomic sequencing, leading to faster and more precise diagnostic outcomes. Similarly, in industries such as aerospace and automotive, it could improve the modeling and simulation of systems, enabling more accurate predictions of failure points and system behaviors [27]. As quantum technology continues to evolve, its integration into diagnostic tools promises to unlock new possibilities in predictive maintenance, fault detection, and system optimization.

7.2 Emerging challenges: cybersecurity and data privacy

As diagnostic technologies become more sophisticated and interconnected, cybersecurity and data privacy emerge as significant challenges. The increasing reliance on IoT devices, cloud computing, and AI for diagnostics has expanded the attack surface for cyber threats, making systems more vulnerable to attacks that could compromise sensitive data or disrupt system operations. Moreover, the collection and analysis of vast amounts of diagnostic data raise concerns about data privacy, particularly in sectors like healthcare, where patient data is involved. Addressing these challenges requires a proactive approach to cybersecurity, incorporating security by design principles, regular security assessments, and the implementation of robust encryption and access control measures. Additionally, compliance with data protection regulations and the adoption of privacy-preserving technologies will be crucial in maintaining trust and ensuring the ethical use of diagnostic data [28, 29].

7.3 Predictions for the next generation of diagnostic tools

The next generation of diagnostic tools is expected to be characterized by greater integration, intelligence, and interconnectivity. AI and ML will become even more embedded in diagnostic processes, not only enhancing the accuracy and efficiency of diagnostics but also enabling predictive and prescriptive analytics. These tools will likely offer more personalized diagnostics and maintenance recommendations, tailored to the specific conditions and usage patterns of each system. Furthermore, the integration of IoT and edge computing will facilitate real-time diagnostics and maintenance across distributed systems, improving responsiveness and reducing downtime [30].

Advancements in sensor technology will also play a critical role, in the development of more sensitive, durable, and energy-efficient sensors that can be integrated into a wider range of systems, including those in remote or harsh environments. The evolution of diagnostic tools will also see a greater emphasis on user experience, with more intuitive interfaces and visualization techniques that democratize access to diagnostic information, making it understandable and actionable for a broader audience.

In addressing emerging challenges, future diagnostic tools will incorporate advanced cybersecurity measures as a fundamental component, ensuring the integrity and confidentiality of diagnostic data. Privacy considerations will also drive the development of diagnostic technologies, with a focus on ensuring that data collection and analysis are conducted in a manner that respects individual privacy and complies with regulatory requirements.

As we look to the future, the continued convergence of technologies such as quantum computing, augmented reality, and blockchain could further enhance the capabilities of diagnostic tools, offering unprecedented levels of accuracy, security, and efficiency in system diagnostics [2]. These advancements promise not only to improve the reliability and performance of systems across industries but also to contribute to the development of more sustainable and resilient infrastructures.

8. Conclusion

8.1 The impact of advanced diagnostics on systems engineering

Advanced diagnostics have fundamentally transformed the landscape of systems engineering, ushering in an era where predictive maintenance, real-time monitoring, and intelligent decision-making have become the norm rather than the exception. This evolution has significantly enhanced the efficiency, safety, and reliability of complex systems across various sectors, including automotive, aerospace, healthcare, and manufacturing. By enabling early detection of potential issues and facilitating timely interventions, advanced diagnostics reduce downtime, extend system lifespan, and optimize performance. Moreover, the integration of AI, ML, and IoT into diagnostics tools has not only improved the precision of fault detection but also personalized maintenance schedules based on system usage and performance data.

The shift toward predictive maintenance models, empowered by advanced diagnostics, represents a paradigm shift in systems engineering, moving away from reactive maintenance strategies toward a more proactive and predictive approach. This transition not only minimizes operational disruptions but also contributes to substantial cost savings and resource optimization. Furthermore, the emphasis on data-driven decision-making enhances the adaptability and resilience of systems, enabling them to respond dynamically to changing conditions and emerging challenges.

8.2 Preparing for future innovations in diagnostic technologies

As the frontier of diagnostics continues to expand, system engineers and industry stakeholders must stay abreast of emerging technologies and trends to harness their full potential. Preparing for future innovations entails fostering a culture of continuous learning and adaptability within organizations, encouraging collaboration across disciplines, and investing in the development of skills and expertise in areas such as AI, quantum computing, and cybersecurity.

The integration of new diagnostic technologies also requires a holistic approach to systems design, considering not only the technical aspects but also ethical, regulatory, and societal implications. Ensuring the security and privacy of diagnostic data, for instance, will be paramount as diagnostics become increasingly interconnected and reliant on cloud-based platforms. Moreover, as diagnostics tools become more sophisticated, there is a growing need for standardization and interoperability to ensure that these tools can be seamlessly integrated into existing systems and processes.

Looking ahead, the future of diagnostics in systems engineering will likely be characterized by even greater levels of automation, intelligence, and integration. By embracing these innovations, engineers can not only enhance the performance and reliability of systems but also contribute to the development of more sustainable, efficient, and resilient infrastructures. Preparing for and adapting to these future trends will be crucial in maintaining the competitiveness and relevance of systems engineering in the face of rapidly evolving technological landscapes.

Acknowledgment

The author acknowledges the use of ChatGPT 4.0 as an AI tool for language polishing of the manuscript.

Conflict of interest

The authors declare no conflict of interest.

Author details

Adeel H. Suhail*, Fiseha M. Guangul and Abdul Nazeer
Middle East College, Muscat, Oman

*Address all correspondence to: adeel@mec.edu.om

References

[1] Russell SJ, Norvig P. Artificial Intelligence: A Modern Approach. Pearson; 2016

[2] Castelvecchi D. Quantum computers are poised to kickstart a second computing revolution. Nature. 2016;**536**:201-204. DOI: 10.1038/536201a

[3] Zhong D, Xia Z, Zhu Y, Duan J. Overview of predictive maintenance based on digital twin technology. Heliyon. 2023;**9**(4):e14534

[4] Zhuang L, Xu A, Wang X-L. A prognostic driven predictive maintenance framework based on Bayesian deep learning. Reliability Engineering & System Safety. 2023;**234**:109181

[5] Madni AM, Sievers M. Systems integration: Key perspectives, experiences, and challenges. Systems Engineering. 2018;**21**(1):82-96. DOI: 10.1002/sys.21412

[6] Pecht M, Jaai R. A prognostics and health management roadmap for information and electronics-rich systems. Microelectronics Reliability. 2010;**50**(3):317-323. DOI: 10.1016/j.microrel.2009.07.002

[7] Lee J, Bagheri B, Kao HA. A cyber-physical systems architecture for industry 4.0-based manufacturing systems. Manufacturing Letters. 2015;**3**:18-23. DOI: 10.1016/j.mfglet.2014.12.001

[8] Kusiak A. Smart manufacturing. International Journal of Production Research. 2018;**56**(1-2):508-517. DOI: 10.1080/00207543.2017.1351644

[9] Johnson SB, Gormley TJ, Kessler SS, Mott CD, Patterson-Hine A, Reichard KM, et al., editors. System Health Management: With Aerospace Applications. John Wiley & Sons; 2011

[10] Lee J, Kao HA, Yang S. Service innovation and smart analytics for industry 4.0 and big data environment. Procedia CIRP. 2014;**16**:3-8. DOI: 10.1016/j.procir.2014.02.001

[11] Dave DM, k. Advancing medical device manufacturing: The convergence of edge computing and industry 5.0. International Journal of Engineering Applied Sciences and Technology. 2023;**8**:126-136

[12] Raghunath K, Vinothkumar V, Venkatesan M, Singh K, Mahesh T, Singh A. XGBoost regression classifier (XRC) model for cyber attack detection and classification using inception V4. Journal of Web Engineering. 2022;**21**:1295-1322

[13] Lei Y, Li N, Guo L, Li N, Yan T, Lin J. Applications of machine learning to machine fault diagnosis: A review and roadmap. Mechanical Systems and Signal Processing. 2020;**138**:106587. DOI: 10.1016/j.ymssp.2020.106587

[14] Ray PP. A survey on internet of things architectures. Journal of King Saud University - Computer and Information Sciences. 2018;**30**(3):291-319. DOI: 10.1016/j.jksuci.2016.10.003

[15] Yang W, Tse PW, Yang X. Machine condition prognosis based on regression trees and one-step-ahead prediction. Mechanical Systems and Signal Processing. 2008;**22**(5):1170-1181. DOI: 10.1016/j.ymssp.2007.11.002

[16] Zhang Z, Zheng Z, Zou Y. Machine learning-based debugging: A survey. IEEE Access. 2020;**8**:142397-142412. DOI: 10.1109/ACCESS.2020.3008139

[17] Garcia-Teodoro P, Diaz-Verdejo J, Maciá-Fernández G, Vázquez E. Anomaly-based network intrusion detection: Techniques, systems and challenges. Computers & Security. 2009;**28**(1-2):18-28. DOI: 10.1016/j.cose.2008.08.003

[18] Schwabacher M, Goebel K. A survey of artificial intelligence for prognostics. In: AAAI Fall Symposium: Artificial Intelligence for Prognostics. Nov 2007. pp. 108-115

[19] Gouriveau R, Medjaher K, Zerhouni N. From Prognostics and Health Systems Management to Predictive Maintenance 1: Monitoring And Prognostics. John Wiley & Sons; 2016

[20] Topol EJ. High-performance medicine: The convergence of human and artificial intelligence. Nature Medicine. 2019;**25**:44-56. DOI: 10.1038/s41591-018-0300-7

[21] Qiu H, Lee J, Lin J, Yu Y. A review of the progressive approach for condition monitoring and fault diagnosis of machine tools. International Journal of Advanced Manufacturing Technology. 2019;**102**:3911-3923. DOI: 10.1007/s00170-019-03363-8

[22] Kusiak A. Smart manufacturing must embrace big data. Nature. 2017;**544**:23-25. DOI: 10.1038/544023a

[23] Lu Y, Morris KC, Frechette S. Current standards landscape for smart manufacturing systems. National Institute of Standards and Technology, NISTIR. 2016;**8107**(3)

[24] Hashem IAT, Yaqoob I, Anuar NB, Mokhtar S, Gani A, Ullah KS. The rise of "big data" on cloud computing: Review and open research issues. Information Systems. 2015;**47**:98-115. DOI: 10.1016/j.is.2014.07.006

[25] Few S. Now you See it: Simple Visualization Techniques for Quantitative Analysis. Analytics Press; 2009

[26] Provost F, Fawcett T. Data Science for Business: What you Need to Know about Data Mining and Data-Analytic Thinking. O'Reilly Media, Inc; 2013

[27] Al-Fuqaha A, Guizani M, Mohammadi M, Aledhari M, Ayyash M. Internet of things: A survey on enabling technologies, protocols, and applications. IEEE Communications Surveys & Tutorials. 2015;**17**(4):2347-2376. DOI: 10.1109/COMST.2015.2444095

[28] Agrawal V, Agrawal S, Bomanwar A, Dubey T, Jaiswal A. Exploring the risks, benefits, advances, and challenges in internet integration in medicine with the advent of 5G technology: A comprehensive review. Cureus. 2023;**15**:e48767

[29] Verma ME, Bridges RA, Sosnowski JJ, Hollifield SC, Iannacone MD. CAN-D: A modular four-step pipeline for comprehensively decoding controller area network data. IEEE Transactions on Vehicular Technology. 2020;**70**:9685-9700

[30] Roman R, Zhou J, Lopez J. On the features and challenges of security and privacy in distributed internet of things. Computer Networks. 2013;**57**(10):2266-2279. DOI: 10.1016/j.comnet.2012.12.018

Chapter 6

Innovation in Complex Systems: Identifying Opportunities with Systems Thinking and Powering Change through Design Thinking

Amin Azad, Farnaz Khodaei, Roya Gazori and Hananneh Ardestani

Abstract

This chapter delves into the dynamic relationship between Systems Thinking and Design Thinking, two influential methodologies reshaping the landscape of innovation. Beginning with an introduction to both approaches, the chapter explores their historical contexts, diverse applications across industries, and prominent models. Design Thinking is elucidated first, highlighting its emphasis on empathetic problem-solving, ideation, and prototyping through tools such as user personas and journey mapping. Meanwhile, systems thinking is characterized by its holistic perspective, focusing on comprehending complex interrelationships and patterns within systems, utilizing tools like causal loop diagrams and system maps for analysis. The chapter proceeds to compare and contrast these two methodologies, ultimately revealing a compelling fusion of their principles. It demonstrates how systems thinking adeptly identifies opportunities within intricate systems, which seamlessly transitions into the realm of Design Thinking for implementing transformative changes and creating ventures. This integrated approach provides a comprehensive framework—leveraging systems thinking for opportunity identification and Design Thinking for impactful implementation—offering a holistic guide for navigating and innovating within complex systems.

Keywords: systems thinking, design thinking, entrepreneurship, innovation, wicked problems

1. Introduction

Amid an era marked by incessant change and an ever-growing complexity of challenges, the need for innovative problem-solving methodologies has never been more pronounced. As we navigate a landscape characterized by multifaceted issues that resist simplistic, one-dimensional solutions, the intersection of Systems Thinking and Design Thinking emerges as a compelling paradigm for addressing the intricacies of our contemporary world.

The imperative to move beyond solutions that are facile and easily implemented into a more nuanced understanding of complex problems finds resonance in the works of Rittel and Webber [1], who introduced the concept of "wicked problems." These are challenges that defy straightforward solutions due to their interconnected nature and the multitude of stakeholders involved.

Recognizing the interdisciplinary nature of contemporary problems, Buchanan [2] advocates for a design practice that goes beyond esthetics, emphasizing its role as a problem-solving tool that engages with complex systems. This perspective aligns with the ethos of systems thinking, which, as discussed by Checkland [3], seeks to understand the interrelated components of a system and their dynamic interactions.

Design Thinking, as popularized by Tim Brown and IDEO, focuses on human-centered problem-solving, emphasizing empathy and iterative prototyping. On the other hand, systems thinking, rooted in the works of Ackoff [4] and Forrester [5], offers a holistic approach to understanding and addressing complex issues by examining the relationships and feedback loops within a system.

The tension between specialization and integration in problem-solving approaches is discussed by Cross [6], who emphasizes the need for both depth and breadth in thinking processes. This tension is particularly evident in the dichotomy between Design and Systems Thinking, where the specialization of tools has been a response to perceived separateness.

Recent market trends also demonstrate this shift in the market and design industry. The design giant, IDEO has strategically laid off roughly 25% of its staff and closed many operations and offices in various locations [7]. This highlights the limitations in some areas of design thinking when addressing complex issues and suggests that systems thinking could be a complementary approach [8].

The role of entrepreneurship in navigating complexity is highlighted by Schumpeter [9], who explains that entrepreneurs are the drivers of disruptive innovation, challenging the status quo. The effectiveness of entrepreneurial thinking in addressing complex problems is further explored by Sarasvathy [10] through the concept of effectuation, highlighting the dynamic and adaptable nature of entrepreneurial decision-making.

As we embark on this exploration, we delve into the historical underpinnings and fundamental principles that define both Design and Systems Thinking. Drawing on the insights of seminal thinkers in these domains, we aim to elucidate the distinct stages and common tools employed by these methodologies, laying the groundwork for a nuanced understanding of their individual contributions to problem-solving.

Building upon this scholarly foundation, we propose an integrative approach—melding the realms of Systems and Design Thinking into a harmonious model. This synthesis aims not to dilute the individual efficacy of these thinking styles but rather to leverage their complementary strengths. Our objective is to present a best-practice model that draws on the scholarly works mentioned, equipping problem solvers with the ability to identify opportunities within complex systems and subsequently execute meaningful, transformative change.

As we delve into the exploration of this integrative model, we seek to bridge the gap between the perceived dichotomy of Design and Systems Thinking, offering a roadmap for practitioners and scholars alike to navigate the intricate terrain of contemporary problem-solving, guided by rigorous academic insights and frameworks.

2. Comparing design and systems thinking

2.1 History of design thinking

Design thinking has a rich and diverse history, with roots tracing back to Peter Rowe's work in architecture at Harvard School of Design [11]. However, its contemporary prominence is largely associated with IDEO and its founder David Kelley, and current CEO Tim Brown [12]. IDEO, initially focused on product development, has expanded its scope to include services, strategies, and even social systems.

Brown advocates for design thinking as a human-centered approach to innovation, emphasizing the importance of understanding people's needs through observation and designing solutions accordingly. He draws examples from innovators like Thomas Edison to illustrate how design thinking can lead to creative and practical solutions, emphasizing multidisciplinary teamwork and user-centered experiences [13].

Herbert Simon, in "The Sciences of the Artificial," views design thinking as a problem-solving tool that engages with complex systems to create efficient and harmonious solutions. He aligns this perspective strongly with systems thinking, stressing the dynamic interactions within systems and their environment [14].

Richard Buchanan's article "Wicked Problems in Design Thinking" further expands on this concept, presenting design thinking as essential for solving complex, ill-defined problems. This perspective has influenced designers, researchers, and specialists across various fields, fostering a multi-dimensional approach to problem-solving (**Figure 1**) [2].

Design thinking was first written about by Peter Rowe, a professor of architecture at the Harvard School of Design. In its current usage, design thinking is more associated with the innovation consultancy firm IDEO, its founder David Kelley, and its current CEO Tim Brown. IDEO, initially focused on product development, has expanded its activities to include the design of services, strategies, and educational and social systems [12].

Brown introduces design thinking as a human-centered approach to the innovation process. This approach emphasizes a deep understanding of people's needs and desires through direct observation and then designing products and services based on this understanding. Brown, using examples from the history of innovation—like the works of Thomas Edison—demonstrates how design thinking can lead to the creation of creative and practical solutions. Brown emphasizes that design thinking goes

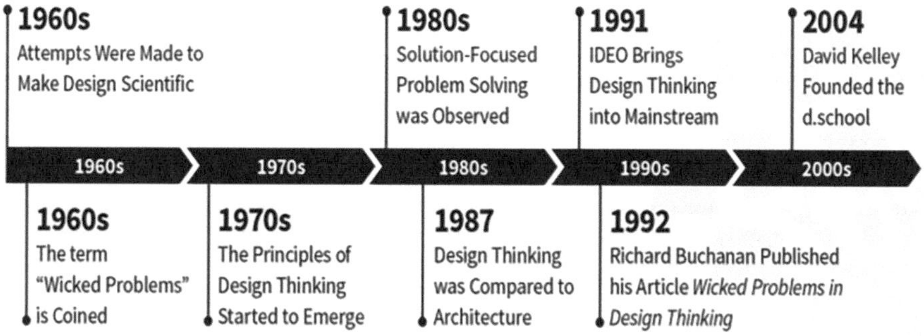

Figure 1.
A visual of the progression of design thinking through time [15].

beyond the creation of physical products and speaks of the importance of multidisciplinary teams and user-centered experiences in the innovation process [13].

Herbert Simon, in his book 'The Sciences of the Artificial,' views design thinking as a tool for problem-solving that goes beyond traditional esthetics and engages with complex systems to create solutions that are efficient and in harmony with their environment. This perspective is strongly aligned with systems thinking and highlights dynamic interactions within a system and between the system and its environment. Also, in this book, Herbert Simon examines design thinking from the perspective of the artificial sciences. He emphasizes the importance of understanding both natural and artificial phenomena and defends a science that considers human goals and purposes in addition to natural laws. Simon explores the concept of "wicked problems" and the need for interdisciplinary approaches in design. Herbert Simon defined design as a science [16].

Richard Buchanan's 1992 article 'Wicked Problems in Design Thinking' has helped to expand and deepen the concept of design thinking, introducing it as a key tool for understanding and solving complex problems that are often ill-defined and not easily solvable. This view has had a widespread impact on designers, researchers, and specialists in various fields, contributing to the formation of a new generation of designers who approach problems with a multi-dimensional and comprehensive perspective [2].

The first Design Thinking Research Symposium, founded by Nigel Cross and Norbert Roozenburg, was held in 1991 at Delft University of Technology. This workshop was organized as an international meeting to discuss emerging research in design thinking, focusing on design cognition and computational modeling of design processes [17].

2.2 Design thinking definition

Design Thinking is defined in various ways, depending on the field that is being studied. These definitions cover various aspects of design thinking, including product innovation, team processes, user focus, and an interactive and experimental approach. A series of significant definitions are described in the following section.

Thomas Lockwood perhaps offers the most accurate definition of design thinking: "A human-centered innovation process that focuses on observation, collaboration, rapid learning, visualization of ideas, rapid prototyping of concepts, and simultaneous business analysis" [18]. Design Thinking is referred to as a field that uses design methods to adapt the needs of people with what is technologically feasible, and a viable business strategy can convert into customer value [13]. Design thinking is therefore recognized as a method or a set of tools for achieving product or service innovation.

In the book 'Change by Design' by Tim Brown, CEO of IDEO, it is emphasized that the fundamental principle of design thinking is to create a deep and empathetic understanding of the challenges that people face. Brown explains that this approach includes understanding complex and mental concepts such as emotions and behavioral motivations, which is different from pure scientific methods that focus more on quantitative research. He introduces design thinking as a third approach, which is a problem-solving approach, rooted in the field of design and aims to create innovative solutions by combining user-focused views with logical analysis and research. (Change by Design).

Design thinking is also defined as a conscious process for obtaining new and relevant solutions that create a positive impact. Design thinking believes in creative abilities and is a process that turns challenges into opportunities [19]. Design thinking, according to Michlewski [20], is defined as a set of methods and tools for achieving product or service innovations. This approach is a conscious process for obtaining new and relevant solutions that create a positive impact, with an emphasis on belief in creative abilities and turning challenges into opportunities. Additionally, design thinking, as defined by Buchanan [2], and Tom Kelley [21], is defined as a specific method for solving complex problems and creating innovative solutions with a user-centric approach and the use of multidisciplinary teams.

Finally, design thinking can be described as a comprehensive approach that both pursues innovation in products and services and, through conscious and creative processes, transforms challenges into new opportunities. This approach, with an emphasis on multidisciplinary collaboration and focus on user needs, turns the complexities present in problems into innovative solutions. The concept of design thinking is proposed as a powerful and effective approach for innovation, generating innovations that go beyond incremental improvements [13] and, in general, a method that leads to better and more competitive business [22–24].

2.3 Industries in which design thinking is used

In an increasingly challenging world, companies require inventive and multifaceted approaches to confront their difficulties, thereby enabling them to achieve victory over their rivals. The implementation of design thinking affords competitive advantage, fosters team collaboration and innovation, enhances sales interactions, facilitates improved product launches, and effectively addresses the tangible requirements of the product [15].

Jeanne Liedtka emphasizes the significance of design thinking in fostering organizational innovation. She asserts that this approach facilitates the identification of new opportunities and enables creative responses to challenges [12]. Buchanan argues that design thinking extends beyond the creation of physical objects or visual representations and instead encompasses the shaping of human experiences and interactions. This broader perspective presents design as both an intellectual and practical endeavor that influences all facets of human life [2].

This approach is not only applicable in educational contexts but also in business and technology, aiding in the development of products and services that directly cater to the needs and desires of users [25]. The utilization of design thinking tools within organizations engenders an experiential learning process that ultimately supports the cultivation of organizational cultures centered on user-centricity, collaboration, risk-taking, and learning.

2.4 Design thinking stages

The field of design thinking has witnessed the contribution of numerous experts who have developed models, each with their distinct ways of enhancing the process's efficacy and strength. These models generally include a range of four to seven stages, thereby reflecting the extensive diversity and profound nature of design thinking in various contexts. The three models discussed below address varying needs, enabling design teams to concentrate on specific project aspects and deliver more efficient solutions.

Liedtka has provided a comprehensive framework for applying design thinking, which includes the four stages. The first step "What is?" involves understanding the current state of the problem or the user's experience. It requires empathizing with the users, defining the problem, and gaining a deep understanding of the challenges and opportunities. The next step, "What if?" is about exploring alternative solutions and generating ideas. It involves ideation and prototyping to create innovative solutions to the defined problem. Subsequently, "What wows?" relates to creating solutions that not only address the problem but also delight the users. It involves refining the ideas and prototypes to ensure that the solutions are compelling and user-friendly. Finally, "What works?" involves testing and implementing the solutions, gathering feedback, and iterating based on the results. It focuses on ensuring that the solutions are viable and effective in addressing the problem [26].

An alternative model from Ambrose and Harris comprises of five sequential phases that can be advantageous to any team utilizing this concept in their process. These phases are as follows: Empathizing, Defining, Ideating, Prototyping, and Testing. In summary, these phases concentrate on comprehending user needs, nurturing creativity, and refining designs through iterative feedback and testing. The model was later matured into a seven-stage process which could be viewed as follows (**Figure 2**) [27].

Another structure presented for applying the stages of design thinking is the "Double Diamond" model. The Diamond Model of Design Thinking is a structured, yet flexible framework that guides the design process across various industries and contexts. It consists of four distinct phases: Discovery, Definition, Development, and Delivery. Each phase plays a crucial role in ensuring that the final product or service is both innovative and aligned with user needs. This model is widely recognized and used by large companies, indicating its effectiveness in fostering creativity and practical implementation in the design process (**Figure 3**) [28].

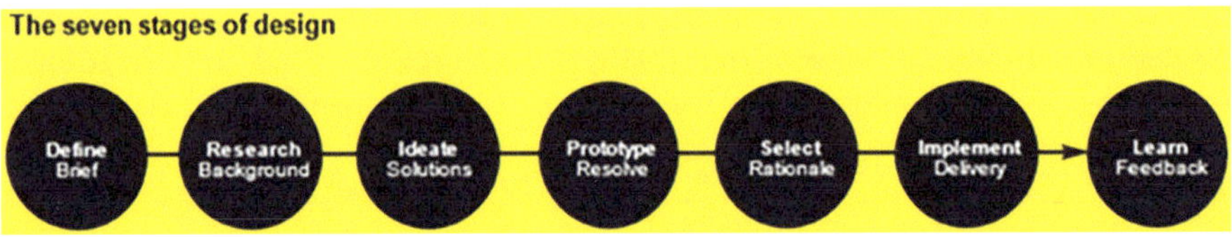

Figure 2.
Seven-stage model of design thinking from Ambrose and Harris [27].

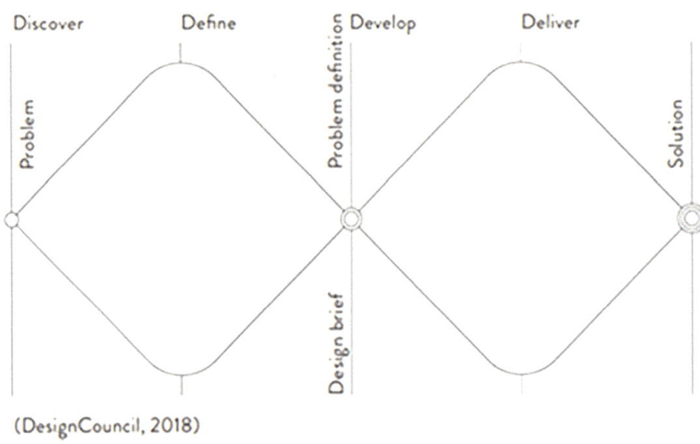

Figure 3.
The double diamond model developed by Gustafsson.

All models prioritize understanding user needs and contexts while each involves ideation to generate solutions. Prototyping and testing are common, emphasizing practical evaluation. Each model provides a different approach tailored to project needs, team dynamics, and goals. While the fifth and seventh stages of design thinking provide a detailed framework for understanding user needs, generating ideas, and implementing solutions, the four-stage model proposed by Liedtka emphasizes a streamlined process that quickly moves from problem understanding to ideation, then focuses on the impact of ideas before testing their viability [26].

2.5 Design thinking tools

The first stage, Empathize, is all about gaining a deep understanding of the users and their needs. Designers employ tools such as observation, interviews, empathy mapping, and user journey mapping to directly observe users in their environment, conduct interviews, and create visual representations of their thoughts, feelings, actions, and needs.

In the Define stage, designers synthesize the information gathered in the empathy stage to define the problem or challenge at hand. Tools like problem framing, persona creation, and stakeholder mapping are used. By clearly defining the problem statement and creating fictional user personas, design teams can align their efforts and consider the perspectives and interests of various stakeholders involved.

The third stage, Ideate, is dedicated to generating a multitude of ideas and potential solutions. Brainstorming, mind mapping, SCAMPER, and "How Might We" statements are commonly employed tools. Through open and free-flowing discussions, visual diagrams, prompt-based idea generation, and reframing challenges into opportunities, designers can explore a wide range of creative possibilities.

Once a diverse set of ideas has been generated, the fourth stage, Prototype, involves bringing those ideas to life through tangible representations. Tools such as paper prototyping, wireframing, storyboarding, and digital prototyping platforms come into play. By creating low-fidelity mockups, skeletal structures, or interactive storyboards, designers can quickly explore and test different concepts, gathering valuable feedback for further refinement.

Finally, in the Test stage, designers gather feedback and insights from users to evaluate and refine their prototypes. Usability testing, interviews, surveys, and iterative feedback techniques are utilized. Through direct observation, user feedback, and quantitative data collection, designers can identify areas for improvement and make iterative adjustments to enhance the user experience [12].

3. Systems thinking

Due to the rapid expansion of technological and economic advancement, enterprises tend to restructure their business models. Consequently, they endeavor to discover novel tools and techniques to address the challenges encountered in large-scale intricate systems [29]. Apart from technical attributes, complex systems possess additional significant characteristics, including ambiguity, emergence, interconnectedness, and socio-economic influences such as employee dynamics, societal factors, and governmental policies. To effectively manage problem areas within complex systems, these features necessitate consideration and attention [30]. One approach toward resolving complex issues involves the utilization of systems thinking. Various perspectives on systems thinking aid in enhancing our comprehension of this methodology. The theory of systems thinking encompasses a collection of concepts, methodologies, and principles that enable us to comprehend phenomena and facilitate future actions [31].

Systems Thinking constitutes an approach to analyzing and addressing real-world predicaments grounded in the fundamental notion of a "system." This discipline enables us to discern how to effectuate change in a manner that promotes the optimal functioning and harmonious integration of systems within the realms of nature and the economy [32]. An alternative viewpoint characterizes systems thinking as "a collection of learning objectives which embody one or more characteristics of systems that generally emanate from systems theory. A set of 'cognitive skills' clarifies that the emphasis lies in the application of systems concepts to problem-solving [33]." Systems Thinking serves as a conceptual framework for contemplating and acquiring knowledge about diverse systems, distinguishing between constituent parts, their interrelations, and the perspectives of different agents. In essence, system thinking encompasses a theory, belief system, and problem-solving approach [34].

3.1 History of systems thinking

Systems Thinking, in comparison to disciplines such as science, theology, education, languages, and history, is a relatively new field. From a more recent perspective, the origins of its various expressions can be traced back thousands of years [35]. In the modern era, the first documented usage of the term "systems thinking" in literature can be attributed to Philip Frank, who expressed a pessimistic view of non-traditional approaches to the study of science in his book Interpretations and Misinterpretations of Modern Physics in 1938 [34]. Von Bertalanffy [36] aimed to unify the sciences under a single framework to address complex systemic problems. In his view, instead of focusing on their differences, scientists from different fields should concentrate on the similarities and connections between them.

According to Arnold [35], Russell Ackoff [37] may have been the first to introduce systems thinking. However, this is not the case, as Ackoff acknowledged the work on systems thinking by Emery [38]. Forster [39], a student of systems thinking since the early 1960s, found that describing systems thinking is both the most challenging and the initial and crucial aspect of the field. Ackoff [37] conceptualized a system of systems, where organizations are considered systems that serve a purpose beyond mere existence or survival. In other words, an organization does not exist solely for the sake of existing, but rather to achieve something meaningful.

Bui conducted a review of Peter Senge's work spanning thirty years [40]. The framework, including The Fifth Discipline [41, 42], and The Dance of Change [43], consists of elements such as personal mastery, mental models, team learning, shared perspectives, and systems thinking. Senge is not the first to discuss systems thinking, as others like [44–46] preceded him. However, Senge made the concept more applicable and accessible in the realms of physics, psychology, business, and management. The figure above, adapted from the article by [47], illustrates a timeline of systems thinking theorists (**Figure 4**).

3.2 Industries in which system thinking is used

Systems Thinking is used extensively in various industries such as engineering, business, management, and healthcare [48]. It is particularly prevalent in the fields

DOI: http://dx.doi.org/10.5772/intechopen.114374

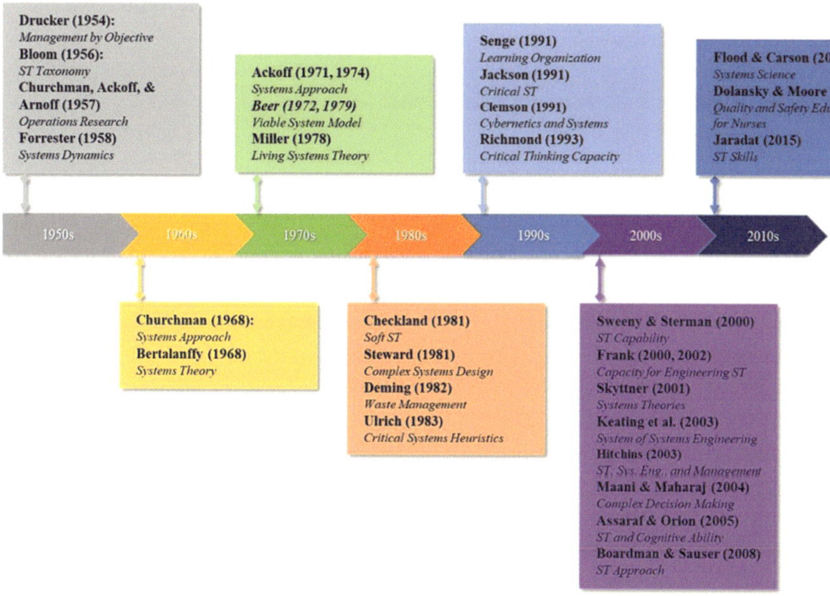

Figure 4.
Timeline of system thinking theorists [47].

of operations, management, and environmental sectors, where it helps in checking for interdependencies, interconnections, and coherence in all aspects of the elements inputted [49]. In healthcare, systems thinking is used to design systems that prevent, catch, or fix natural human slips to avoid errors [50]. It is also used to make healthcare safer by implementing tools like barcode scanning and enteral feeding connectors. In engineering [51], systems thinking enables the study and understanding of interactions between individuals, departments, and business units within an organization or overall system-of-systems design. Overall, systems thinking is a valuable approach used in various industries to analyze and improve complex systems.

3.3 The importance and application of systemic thinking

According to the scholarly work of [52] on the significance and application of system thinking, it can be contended that the utilization of systems thinking broadens our cognitive horizons and facilitates the articulation of problems in novel and distinct manners, thereby expanding the array of available alternatives for resolving a given issue. Concurrently, the principles underpinning systems thinking enlighten us on the absence of flawless solutions. The decisions we make inevitably exert an influence on other components within the system. By preemptively acknowledging the ramifications of each transaction, we can mitigate its severity or even employ it to our advantage. Ergo, systems thinking empowers us to make informed choices.

Moreover, systems thinking bestows immense value in constructing compelling narratives that elucidate the mechanics of a system. For instance, the exercise of causal loop diagramming compels a team to collectively generate shared visual representations or narratives that depict a particular scenario. These tools serve as efficacious means for identifying, elucidating, and effectively communicating our comprehension of systems, particularly in group settings.

Problems that are most amenable to systems thinking intervention possess the following distinctive attributes:

- The subject matter is of considerable import.
- The problem at hand is chronic in nature, rather than a one-off occurrence.
- The problem is familiar and possesses a documented history.
- Prior attempts at resolving this predicament have proven unsuccessful.

3.4 Systems thinking stages

There exist varied methodologies concerning the phases of systems thinking. In the subsequent discourse, we shall expound upon several of these methods, and subsequently, we shall acquaint ourselves with the cycle that we aim to employ in this scholarly article. According to [53], the stages of system dynamics can be delineated by the subsequent cycle (**Figure 5**).

Michael Goodman presents a framework consisting of six sequential steps that can be employed when addressing intricate problems utilizing systems thinking principles [54]. These steps encompass the following elements: storytelling, visual representation of behavior over time through graphs, formulation of a concise and targeted statement, recognition and analysis of underlying structures, identification of the forces that impact the ecosystem and its various functions, and the development of a comprehensive problem-solving plan.

Barry Richmond asserts that the process of system thinking encompasses a sequence of four distinct stages [55]. During the execution of this process, a total of seven skills associated with systems thinking are effectively employed, as demonstrated (**Figure 6**).

In the subsequent section, the model outlined beneath pertains to [56, 57] and this methodology is employed in this scholarly piece for the phases of systemic thinking (**Figure 7**).

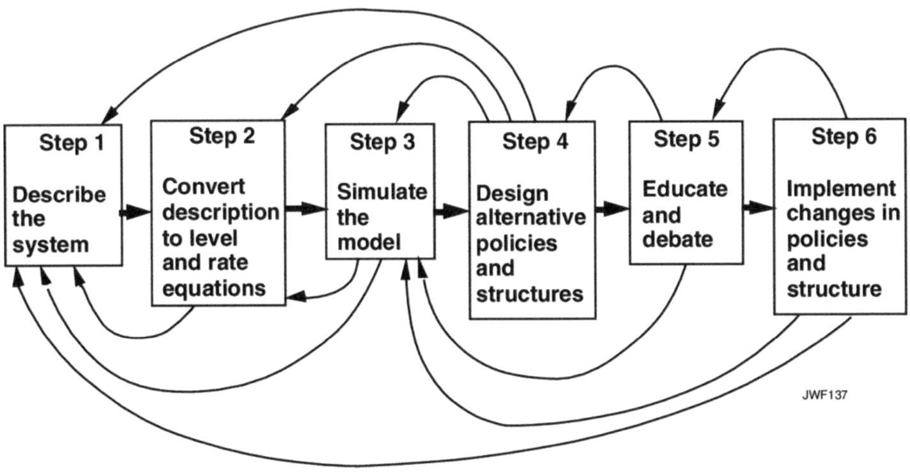

Figure 5.
Stages of system dynamics.

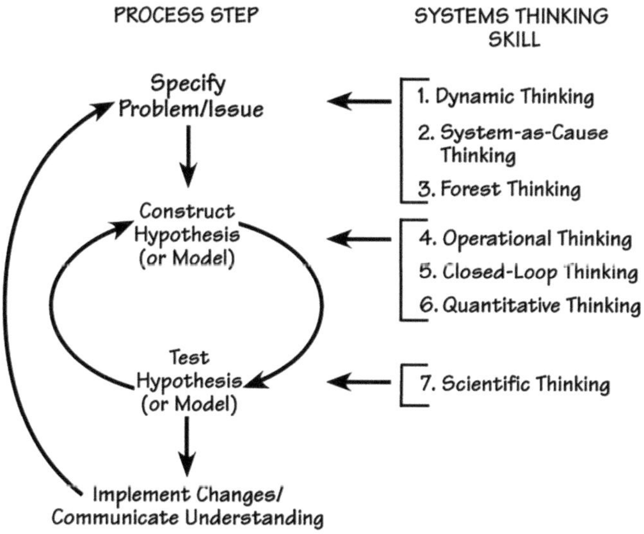

Figure 6.
Process of system thinking.

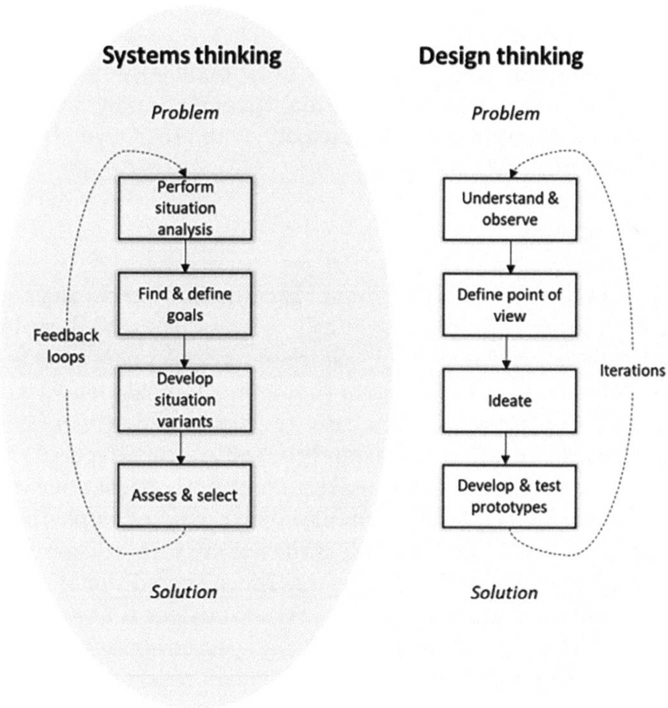

Figure 7.
Convergence between the system thinking and design thinking approach.

Based on the model, to implement the principles of system thinking in a systematic manner, it is necessary to proceed as follows.

During the first phase, a comprehensive analysis of the current situation must be conducted. This entails examining novel ideas or emerging technologies and their potential impact on the capabilities, opportunities, and needs of the organization and its stakeholders. In this initial stage, fundamental concepts are explored through the evaluation of technological risks and the readiness levels of the technology. The focus lies on the problem at hand, as well as the potential solutions, what is feasible, and what is not.

The second phase entails the identification and definition of the desired objective. Systems thinking places great importance on the identification of business or mission requirements, as well as the needs of the stakeholders. Additionally, careful consideration must be given to estimating the costs and planning for the development process.

The third phase involves the development of various types of positions or concepts. Systems Thinking suggests the identification of multiple candidates for a given concept that could potentially offer new capabilities, enhance overall performance, or reduce costs for stakeholders. During this stage, the use of mockups can prove valuable in comprehending and preparing for the subsequent testing phase.

Finally, the fourth phase revolves around the evaluation and selection of candidate concepts. Systems Thinking emphasizes the necessity of testing and assessing potential concepts. Models and prototypes are considered indispensable in conducting feasibility studies, as well as gaining a deeper understanding of stakeholder needs, making architectural trade-offs, and identifying risks and opportunities. The output of the conceptual phase in systems thinking is a clearer comprehension of the business or mission requirements and stakeholder needs and aligning these needs with the expectations of the stakeholders.

Considering the slight differences that exist in different approaches to the stages of system thinking, however, in all cases, the main goal is to know the components of the system, connect the components with each other, and then try to find the problem and take steps to solve the problem.

3.5 Systems thinking tools

System thinking tools are used to understand complex systems and their dynamics, and they enhance problem-solving processes. These tools include causal loop diagrams, behavior over time graphs, stock-flow diagrams, and computer models. They are used at different stages of problem-solving depending on the complexity of the problem. For example, in a case study on emergency medical services, a causal loop diagram was used to understand the causes and effects of rising EMS demand [58]. This tool helped decision-makers identify the most relevant interventions and prioritize potential interventions based on predicted implementation difficulty and effect [59]. Another example is the use of system diagnostic processes to build understanding of a system. This process involves traditional data analysis, systems thinking methodologies, and frameworks to analyze the system and identify key issues and interventions [60] the article by [61] presents a compiled classification of systems thinking tools. This classification encompasses various use cases of these tools.

Most systems thinking tools mentioned in the literature primarily aim to facilitate the comprehension of a system and the co-designing of solutions. However, there is a noticeable dearth of tools that foster dialog, collaboration, and the assessment of progress, including implementation, monitoring, and evaluation. Among the publications reviewed, 19 discussed a general system thinking approach without specifying any tools, while 54 articles provided insights into one or more tools. The subsequent section provides a detailed description of the tools that were reported.

Tools for understanding the system encompass several distinct categories. These include tools for visualizing situations, such as causal loop diagrams (CLD), system mapping, concept maps, system archetypes, rich images, Bayesian belief networks, social network analysis, influence diagrams, and Sankey diagrams. Additionally, tools for strategic thinking involve system archetypes, soft systems methodology, scenario and vision development, social network analysis, and PQR (what, how, why) statements.

In terms of facilitating dialog and collaboration, there are specific tools that aid in stakeholder analysis, multi-stakeholder dialog, and interactive management.

Lastly, collaborative design tools for generating solutions encompass collaborative modeling, group model building, and stock and flow diagrams.

4. Comparing design and systems thinking

In a broad sense, when it comes to examining and contrasting the origins and foundations of each of these respective disciplines, it is conceivable to assert that if we are to undertake a comparative analysis, a fundamental observation can be made: The origins of design thinking can be traced back to the mid-twentieth century, specifically the 1950s and 1960s when it first started gaining prominence in the fields of architecture and engineering. Since then, design thinking has evolved and established itself as a dominant methodology for fostering innovation. This approach seamlessly integrates various aspects, including human factors, technology, and strategic considerations, to effectively address complex challenges that have developed over the course of several decades. On the other hand, systems thinking takes a different approach by emphasizing an understanding of entire systems rather than focusing solely on individual components. This holistic perspective makes it particularly well-suited for tackling intricate and multifaceted systems that have been purposefully designed, such as the healthcare system or urban environments. By considering the interconnectedness and interdependencies within these systems, systems thinking offers a comprehensive framework for analysis and problem-solving.

If we desire to further investigate and analyze these two distinct forms of cognition with regards to the objectives, they each strive to achieve, it becomes evident that a deeper comprehension will be gained design thinking, as a problem-solving approach, aims to create innovative and human-centered solutions to specific challenges. Design thinking recognizes the need for creative and multifaceted solutions to navigate complex and uncertain challenges. It encourages designers to understand the problem, and think outside the box, fostering the generation of new and innovative ideas to address specific problems. Also, design thinking places a strong emphasis on understanding and empathizing with users' needs and desires. By adopting a human-centered perspective, design thinkers aim to develop solutions that enhance user experiences and meet their specific requirements. Furthermore, design thinking is recognized as a powerful approach for encouraging innovation within organizations. It helps managers and practitioners develop innovative strategies and find new opportunities for growth by challenging existing assumptions and exploring novel possibilities.

The goals of design thinking and systems thinking differ in their emphasis and objectives. Systems Thinking aims to understand and address complex problems by adopting a holistic perspective, expanding thinking, anticipating impacts, and making informed choices. On the other hand, design thinking focuses on creative problem-solving, user-centricity, innovation, and considering the fit of solutions within the larger system. These complementary approaches offer unique benefits and can be effectively combined to address complex challenges in various domains. Understanding their distinct goals enhances the problem-solving toolkit and enables more comprehensive and effective solutions.

If we want to compare the applications of system thinking and design thinking in different industries, we must say that, design thinking is frequently utilized in various domains such as product development, service provision, and user experience (UX) design. These applications highlight the versatility and applicability of design thinking as a problem-solving approach. Additionally, design thinking is instrumental in addressing challenges that are inherently tackled by designers. By embracing systems principles, design thinking has consistently fostered the creation of sustainable designs. Conversely, systems thinking is highly regarded both internally within organizations and externally across various sectors. This approach is essential for comprehending and designing intricate, multifaceted, and dynamic systems that are tailored toward specific purposes, such as healthcare and social issues. The complexity associated with problem identification in these domains necessitates a system thinking perspective. Systems Thinking proves to be invaluable in the planning and execution of portfolio strategies within large-scale product organizations. This approach aids in unraveling the intricacies and interdependencies that exist within the system, thereby facilitating effective decision-making and implementation.

If we desire to engage in a comprehensive analysis of the processes involved in utilizing the tools of design thinking in a comparative manner with the concept of systems thinking, we should know that both system thinking and design thinking emphasize understanding the problem context, stakeholder needs, and project objectives. Systems Thinking emphasizes analyzing technology and risks, while design thinking focuses on user needs and creative problem-solving. Design thinking places a strong emphasis on creative ideation and generating multiple options to solve the problem. Both approaches recognize the importance of prototyping and testing ideas. Design thinking emphasizes iterative prototyping and gathering user feedback, while systems thinking uses prototypes for feasibility studies and risk identification. Systems Thinking considers stakeholder needs and aligns them with project goals. Design thinking involves collaboration with multidisciplinary experts and user involvement throughout the process. System thinking can be iterative but is often seen as a more linear process with distinct stages. Design thinking is often described as a cyclical and iterative process, where steps may occur simultaneously or with varying depth.

In summary, both system thinking and design thinking share common elements such as problem understanding, prototyping, and stakeholder engagement. However, design thinking emphasizes creative ideation, user-centricity, and iterative prototyping, while systems thinking focuses on analyzing technology and aligning stakeholder needs with project objectives. The two approaches can be complementary and used together to address complex challenges effectively.

Both systems thinking and design thinking employ a variety of tools that aid in understanding complex systems and generating innovative solutions. However, the focus and application of the tools are different. Systems Thinking tools primarily focus on understanding the system itself, its components, and their relationships. They help identify patterns, analyze causality, and visualize the structure and dynamics of the system. Design thinking tools are more customer-centric and focus on understanding user needs, improving the customer experience, and generating creative ideas for solutions. They emphasize visualization, empathy, and rapid prototyping. While some tools may overlap between the two approaches, the overall emphasis and objectives of the tools differ. System thinking tools aim to analyze and understand complex systems, while design thinking tools aim to empathize with users, generate ideas, and prototype solutions to improve the customer experience. Both approaches can be complementary, and the tools from each can be used interchangeably depending on the context and problem at hand.

5. An alternative framework

In navigating the complex terrain of organizational change, it is imperative to recognize that effective transformation requires more than just addressing the surface-level symptoms. As highlighted by John Sterman in "Systems Thinking and Modeling for a Complex World" [62], altering a complex system often introduces unforeseen challenges. In fact, Sterman advocates for a cautious approach, emphasizing that the best solution may be to delay making changes due to the potential for worsening consequences.

A comprehensive understanding of the intricacies of a system and the users within it is fundamental for creating meaningful change, whether in a simple or highly complex organization. The complexity of a system necessitates an approach that goes beyond merely fixing the immediate issue at hand. Rather, it demands an exploration of the interconnectedness and root causes within the system.

Enter systems thinking—an approach that enables us to delve into the depths of a challenge, understand its facets, and unravel the intricate web of relationships within the system. Yet, understanding the problem in its entirety is only one side of the coin. The ideal solution not only comprehends all aspects of the issue but also considers how it impacts the diverse stakeholders involved. This is where design thinking comes into play, tailoring solutions to align with the unique circumstances and needs of the users.

Having identified the problem and crafted a customized solution, the critical phase of implementation awaits. The efficacy of a solution is only as good as its integration into the system and the value it brings. It is at this juncture that we revisit system thinking, implementing the changes into the system while meticulously mapping and analyzing the consequences of each modification.

This cyclical and iterative process forms the basis of our proposed model—an interplay between Systems Thinking and Design Thinking, operating as a feedback loop. Each iteration refines the solution, considering the dynamic nature of complex systems. Illustrated in **Figure 8**, the proposed model emphasizes the ongoing dialog between these two methodologies, recognizing that stability is achieved through a series of iterations and refinements.

In conclusion, the fusion of Systems Thinking and Design Thinking offers a holistic and adaptable approach to organizational change. By embracing the feedback loop model, organizations can navigate the complexities of their systems, understand the nuanced needs of users, and iteratively refine solutions until stability is achieved. This integrative model stands as a testament to the dynamic and interconnected nature of problem-solving, providing a robust framework for creating sustained and effective change within complex organizational ecosystems.

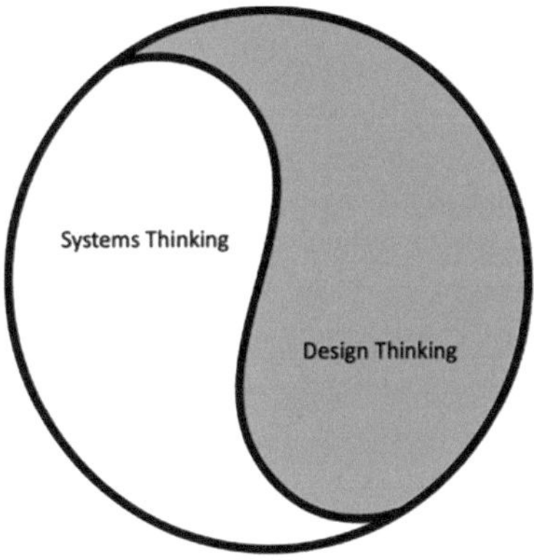

Figure 8.
A proposed methodology for an iterative process between systems and design thinking for complex problem-solving and change implementation.

Acknowledgements

The authors express their gratitude for the invaluable guidance and supervision provided by esteemed supervisors, including Dr. Kambiz Talebi.

Author details

Amin Azad[1*], Farnaz Khodaei[2], Roya Gazori[2] and Hananneh Ardestani[2]

1 University of Toronto, Toronto, Canada

2 University of Tehran, Tehran, Iran

*Address all correspondence to: amin.azadarmaki@mail.utoronto.ca

References

[1] Rittel HWJ, Webber MM. Dilemmas in a general theory of planning. Policy Sciences. 1973;**4**(2):155-169

[2] Buchanan R. Wicked problems in design thinking. Design Issues. 1992;**8**(2):5-21

[3] Checkland P. Systems Thinking, Systems Practice. John Wiley & Sons; 1999

[4] Ackoff RL. Creating the Corporate Future: Plan or be Planned for. Wiley; 1981

[5] Forrester JW. Industrial Dynamics. MIT Press; 1961

[6] Cross N. Designerly Ways of Knowing. Springer; 2007

[7] Wilson M. Design giant Ideo cuts a third of staff and closes offices as the era of design thinking ends [Internet]. FastCompany. 2023. Available from: https://www.fastcompany.com/90976682/design-giant-ideo-cuts-a-third-of-staff-and-closes-offices-as-the-era-of-design-thinking-ends

[8] Blois M, Coninck P. The Dynamics of Actors' and Stakeholders' Participation: An Approach of Management by Design. Architectural Engineering and Design Management. 2008;**4**:176-188. DOI: 10.3763/AEDM.2008.0097

[9] Schumpeter JA. The Theory of Economic Development: An Inquiry into Profits, Capital, Credit, Interest, and the Business Cycle. Transaction publishers; 1934

[10] Sarasvathy SD. Causation and effectuation: Toward a theoretical shift from economic inevitability to entrepreneurial contingency. Academy of Management Review. 2001;**26**(2):243-263

[11] Krūmiņa A. Design thinking in the context of pedagogy. 2018;**4**:423-434. DOI: 10.17770/SIE2018VOL1.3300

[12] Liedtka J, King A, Bennett K. Solving problems with design thinking: Ten stories of what works. Columbia University Press; 2013

[13] Brown T. Design thinking. Harvard business review. 2008;**86**(6):84

[14] You X, Hands D. A Reflection upon Herbert Simon's Vision of Design in The Sciences of the Artificial. The Design Journal. 2019;**22**(sup1):1345-1356. DOI: 10.1080/14606925.2019.1594961

[15] Reddy ENK, Reddy NS. Design thinking – Evolution & its Importance in business. Interantional Journal of Scientific Research in Engineering and Management. 2023;**07**(03):1-6. DOI: 10.55041/ijsrem18122

[16] Simon HA. The Sciences of the Artificial. (n.d)

[17] Cross N. A brief history of the design thinking research symposium series. Design Studies. 2018;**57**:160-164

[18] Lockwood T. Design thinking: Integrating innovation, customer experience, and brand value. Simon and Schuster; 2010

[19] IDEO. Design Thinking for Educators Toolkit. 2012. Retrieved from: http://designthinkingforeducators.com/

[20] Michlewski K. Design Attitude. 1st ed. Routledge; 2015. DOI: 10.4324/9781315576589

[21] Kelley T. Prototyping is the shorthand of innovation. Design Management Journal (Former Series). 2001;**12**(3):35-42

[22] Clark K, Smith R. Unleashing the power of design thinking. Design Management Review. 2008;**19**(3)

[23] Boland R, Collopy F, editors. Managing as designing (p. 298). Stanford, CA: Stanford Business Books; 2004

[24] Martin RL. The design of business: Why design thinking is the next competitive advantage. Harvard Business Press; 2009

[25] Razzouk R, Shute V. What is design thinking and why is it important? Review of Educational Research. 2012;**82**(3):330-348

[26] Liedtka J. Learning to use design thinking tools for successful innovation. Strategy and Leadership. 2011;**39**(5):13-19

[27] Ambrose G, Harris P. Basics Design Basics Design 08: Design Thinking. 2010. Available from: https://books.google.com/books?id=9klpFfZDnWgC&pgis=1

[28] Gustafsson D. Analysing the Double Diamond Design Process through Research & Implementation 2 Content. 2019. Available from: https://core.ac.uk/download/pdf/224802861.pdf

[29] Bankes SC. Tools and techniques for developing policies for complex and uncertain systems. Proceedings of the National Academy of Sciences. 2002;**99**(suppl_3):7263-7266

[30] Jaradat RM, Keating CB, Bradley JM. Individual capacity and organizational competency for systems thinking. IEEE Systems Journal. 2017;**12**(2)1203-12108

[31] Rajagopalan R. Immersive Systemic Knowing: Advancing Systems Thinking beyond Rational Analysis. Springer Nature; 2020. Available from: https://link.springer.com/book/10.1007/978-3-030-49135-2

[32] Arnold RD, Wade JP. A definition of systems thinking: A systems approach. Procedia—Procedia Computer Science. 2015;**44**:669-678. DOI: 10.1016/j.procs.2015.03.050

[33] Verhoeff RP, Knippels MPJ, Gilissen MGR. The theoretical nature of systems thinking. Perspectives on Systems Thinking in Biology Education. 2018;**3**(June):1-11. DOI: 10.3389/feduc.2018.00040

[34] Cabrera D, Cabrera LL, Midgley G. The four waves of systems thinking. Journal of Systems Thinking. 2023:1-51

[35] Arnold RD. Systems Thinking: Definition, Skills, Simulation, and Assessment. United States: Stevens Institute of Technology; 2021

[36] Von Bertalanffy L. Problems of general system theory. Human biology. 1951;**23**(4):302

[37] Ackoff RL. Towards a system of systems concepts. Management science. 1971;**17**(11):661-671

[38] Emery WB. National and International Systems of Broadcasting; Their History, Operation and Control. 1969

[39] Forester J. Bicycle transportation: a handbook for cycling transportation engineers. Mit Press; 1994

[40] Bui HT, Galanou E. Translation of systems thinking to organizational goals: A systematic review. Journal of General Management. 2022;**47**(4):233-245

[41] Senge P. The Fifth Discipline: The Art and Practice of the Learning Organization. New York: Doubleday/Currency; 1990a

[42] Senge P. Peter Senge and the learning organization. Dimension. 1990b:14

[43] Senge P, Kleiner A, Roberts C, Ross R, Roth G, Smith B, Guman EC. The dance of change: The challenges to sustaining momentum in learning organizations. 1999

[44] Ackoff RL. The Systems Revolution. Long Range Planning. 1974;7(6):2-20

[45] Ackoff RL. O. R. and the systems movement: Mappings and conflicts. 1983;**34**(8):661-675

[46] Churchman CW. The design of inquiring system. NY: Basic Books. 1971

[47] Hossain NUI, Dayarathna VL, Nagahi M, Jaradat R. Systems thinking: A review and bibliometric analysis. Systems. 2020;**8**(3):23

[48] Chan W. The Role of Systems Thinking in Systems Engineering, Design and Management. Clinical and Experimental Dermatology. 2015;**17**:126-132. DOI: 10.9744/CED.17.3.126-132

[49] Khan S, Madnick S. Protecting Chiller Systems from Cyberattack Using a Systems Thinking Approach. Network. 2022;**2**(4):606-627

[50] Roxas FMY, Rivera JPR, Gutierrez ELM. Bootstrapping tourism post-COVID-19: A systems thinking approach. 2022;**22**(1):86-101. DOI: 10.1177/14673584211038859

[51] Dugan KE, Mosyjowski EA, Daly, SR, Lattuca LR. Systems thinking assessments in engineering: A systematic literature review. Systems Research and Behavioral Science. 2022;**39**(4):840-866

[52] Goodman M. Systems Thinking as a Language. The systems thinker; 1991

[53] Forrester JW.System dynamics, systems thinking, and soft OR. System dynamics review. 1994;**10**(2-3):245-256

[54] Goodman M, Karash R. Six steps to thinking systemically. The systems thinker. 1995;**6**(2):16-18

[55] Richmond B, Peterson S. An introduction to systems thinking. Lebanon, NH: High Performance Systems. Incorporated. 2001

[56] Lewrick M, Link P, Leifer L. The Design Thinking Playbook: Mindful Digital Transformation of Teams, Products, Services, Businesses and Ecosystems. John Wiley & Sons; 2018

[57] Lewrick M, Link P, Leifer L. The design thinking toolbox: A guide to mastering the most popular and valuable innovation methods. John Wiley & Sons; 2020

[58] Fisher DM, Systems Thinking Association. Systems thinking activities used in K-12 for up to two decades. In: Frontiers in Education. Vol. 8. Frontiers Media SA. Feb 2023. p. 1059733. DOI: 10.3389/feduc.2023.1059733

[59] Rehbock C, Krafft T, Sommer A, Beumer C, Beckers SK, Thate S. Systems thinking methods: A worked example of supporting emergency medical services decision—Makers to prioritize and contextually analyse potential interventions and their implementation. Health Research Policy and Systems. 2023;**21**(1):42. DOI: 10.1186/s12961-023-00982-y

[60] Walls E, Savage L. Can Systems Thinking Tools Help us Better Understand Education Problems and Design Appropriate Support ? Reflections on a Test Case. In Systems Thinking in International Education and Development. Edward Elgar Publishing; 2023. pp. 204-221

[61] Nguyen LKN, Kumar C, Jiang B, Zimmermann N. Implementation of systems thinking in public policy: A systematic review. Systems. 2023;**11**(2):64

[62] Sterman J. Business dynamics: Systems thinking and modelling for a complex world. 2000

Chapter 7

Perspective Chapter: Chip I/O Design Fundamentals, Methodologies and Challenges

Ikechi A. Ukaegbu and Arjuna Marzuki

Abstract

The dramatic increase in processing power due to the combined effects of integrated circuit scaling and shifts in computer architectures from single core to multi-core systems has led to the highly scaled on-chip aggregate bandwidths to the Tb/s range. Consequently, this has resulted to the increase in the amount of data transfer between chips to increase overall system performance. However, due to the limited I/O pin count in chip packages and PCBs, designing efficient high-speed and low power chip I/O is important. This chapter proposes a configurable I/O driver that can be configured between two single-ended voltage-mode drivers, a differential voltage-mode driver and a differential current-mode driver to address electronic signaling mismatches between integrated circuits for 0.4–8.0Gbps applications. Four control schemes, namely, impedance, output swing, slew rate and de-emphasis control scheme, were integrated into the configurable I/O driver to improve signal integrity by addressing the ISI, output switching noise, reflections and output swing deviation. Through the supplementary control from the four digital schemes, the proposed configurable I/O driver is capable of supporting multiple signaling types while improving high-speed signal integrity, thereby enhancing the performance of computer systems for better consumer experiences.

Keywords: chip input/output, signal/power integrity (SI/PI), noise, pin count, EMC

1. Introduction

Microprocessors in modern computer and computing systems can be coupled to various interfaces and the external world through the chip input/output (I/O) modules. Over the years, the performance of modern computer systems has seen intense improvements due to the advancements in semiconductor process. This has allowed the continuous increase in the number of transistors in a single integrated circuit (IC), according to Moore's Law [1]. Subsequently, the improvement of processor performance in terms of speed and power has also dramatically increased due to the merit of larger transistor budget and the faster transistor switching speed [2]. Moreover, multi-core architecture is a new trend of processor design to further improve processor performance through parallel tasks performed by the cores. The chips of these cores can be placed on a common substrate through MCM (Multi-chip Module) packaging technology [3–6]. There are three main MCM technologies: MCM- C, which uses co-fired multilayer ceramic substrate technology; MCM-D, which uses thin-film deposition technology and MCM-L, which uses lamination technology [7]. Among these main MCM technologies, MCM-L is widely used in microprocessor packaging industries for high volume consumer and low cost products [6]. Therefore, the processor performance can be further boosted with multi-core architecture realized by MCM-L packaging technology.

Unfortunately, the performance improvements in the microprocessors has not lead to comparable improvement in the performance of computer and computing systems [8], which is primarily due to the chip I/Os [3]. This is because microprocessor and application specific integrated circuit (ASIC) vendors provide I/O modules with fixed output driver circuits that are designed to compromise/adapt to various system applications [9]. Each ASIC vendor has a different cost-performance-risk-schedule trade-off for the combination of application, chip supplier, chip assembler, substrate fabricator, module assembler and end user [10]. Hence, depending on the decision of vendors, it becomes challenging for a chip I/O with fixed output driver circuits to achieve optimum performance across a variety of applications [7]. Any signaling mismatches between the processor and the chip I/O would clearly impact the overall system performance, resulting in delays that limit the computer system from achieving higher/expected computing speed [11]. A non-optimal solution is to reduce the application operating speed by reducing the speed of processor to an extent that the receive signal eye opening is just sufficient for data detection and recovery [11]. Although the system requirements still can be achieved with lower processor speed, but the system performance is not at best optimized due to the compromise of speed between the chip I/O and the processor. Also, as the operating frequency of ASIC application continue to increase, signal integrity on data transmission between chip I/Os has become a bottleneck due to several types of noise, including Inter-Symbol Interference (ISI) noise, switching noise, and reflection noise [8, 12–14]. ISI is caused by frequency dependent attenuation of transmission channel that increases with operating frequencies and degrades noise margins, giving small eye opening and resulting in very low possibility of correct data detection and recovery [15–18]. Moreover, output drivers in advanced technologies can switch faster due to lower threshold voltage, V_T, but this results a considerable supply current change in a short time (di/dt) and causes switching noise on power supply and output voltage of drivers [19–21]. Also, reflection noise occurs when the output impedance of output drivers is mismatched with characteristic impedance of transmission channel at high frequencies [22, 23]. Additionally, the trend in scaling down CMOS process technology dramatically deteriorates the channel-length modulation effect and this causes challenges in designing I/O driver circuits that employs current-mode transmission like current-mode drivers. Channel length modulation causes symmetric error between current mirror and current-mode driver, resulting output swing that has large deviations from the desired value [15]. Moreover, the ISI, switching noise, and reflection noise can further deteriorate the output swing deviations, thereby reducing the signal eye opening and causing difficulties for correct data detection [24].

Tradeoffs have been made between electronic signaling and operating frequency to ensure that the computer system can function. However, the system performance is not at best optimized due to the compromise of speed between I/O modules and processor [9]. Thus, configurable input/output (I/O) driver is desired whereby different I/O driver types can be configured from the I/O driver to support multiple signaling types without compromising system performance across a variety of applications [25]. In addition to signal integrity issues at high frequencies due to technology scaling, channel length modulation causes symmetric error between current mirror and current-mode driver, resulting to an output swing that has large deviations from the desired value [15]. These output swing deviations can be further deteriorated by ISI, switching noise and reflection noise, thereby degrading signal integrity and causing difficulties for correct data detection [24]. Also, the general trend in chip size reduction due to transistor technology scaling has led to chip size reduction and the advancement in packaging technology. However, due to the limited I/O pin count in these chip packages and PCBs, designing efficient high-speed and low power chip I/O is important.

Hence, in this chapter, we present the design of a configurable I/O driver circuit that can be configured between voltage-mode and current-mode driver to address electronic signaling mismatch between chips and achieve optimum performance across different processor-intended applications. The configurable I/O driver supports a wide range of data-rate of 0.4–8.0 GHz. Also, in the design, four control schemes, namely, impedance control, output swing control, slew rate control and de-emphasis were implemented in the configurable I/O driver to address ISI, output switching noise, reflections and output swing variation. Thus, improving signal integrity at high data-rates. Clear eye openings with improved jitter performance were obtained at high data-rates. This work is conducted in a 45 nm CMOS technology.

2. Chip I/O design overview

This section discusses the overview of several types of output driver, their operation, pros, and cons. In addition, several design techniques to improve driver circuit performance are discussed, including differential signaling; slew rate control on output; output impedance control; and de-emphasis technique to compensation the ISI impacts.

2.1 Output circuit driver

An output driver encodes a symbol into a current or voltage form for transmission over a line. An output driver can also be referred to as a transmitter. In this chapter, the term "output driver" and "transmitter" is used interchangeably. A good transmitter has output impedance optimized for selected transmission mode. Voltage-mode driver and current-mode driver are two fundamental output driver circuits that will be discussed in this section.

2.1.1 Voltage-mode drivers

A voltage-mode transmitter has very small output impedance due to its implementation using large size transistor that operates in the linear region [20]. The simplest circuit of voltage-mode driver is a push-pull driver implemented in a digital CMOS inverter. **Figure 1a** shows the architecture of push-pull driver that was very popular in the early 1980s due to simple implementation and low power consumption [8]. When the gate voltage VIN is low, the NMOS is turned-off, while PMOS is turned-on to pull the output voltage up to VDD and vice versa. **Figure 1b** depicts the I–V characteristic of a NMOS when the gate voltage of push-pull driver is high.

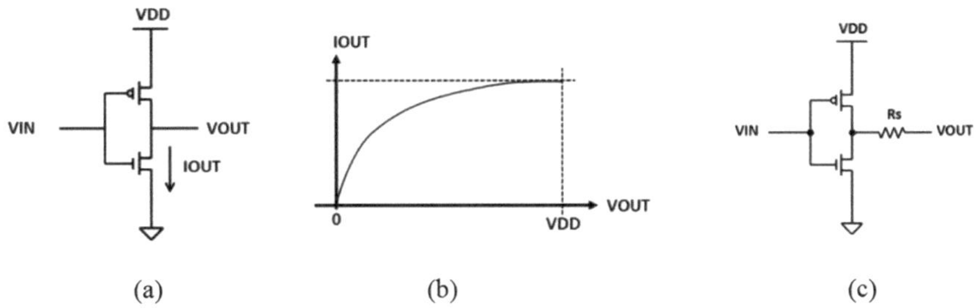

Figure 1.
(a) Push-pull driver circuit, (b) non-linear I–V curve of push-pull driver [8], and (c) SST driver circuit with series resistor at output node [15].

However, the non-linear I–V characteristic in combination with small output impedance of voltage-mode driver are mismatched with the characteristic impedance of transmission channel, thus causing it to suffer from signal reflection, especially when the operating frequency is increasing in modern IC applications [22].

Therefore, another voltage-mode driver called Source-Series Terminated (SST) driver in **Figure 1c** can solve the reflection issues encountered by push-pull driver [15]. The basic structure and operation is the same as that in the push-pull driver, the series resistor, R_S, is connected to the node between pull-up and pull-down transistor. This is an essential supplementary element for an output driver operating in high-speed applications, as the output driver not only transmits data, but can also act as self-terminator to minimize any reflected signals [26]. The series resistor value is often chosen close to transmission channel impedance to minimize signal reflections. However, this requires large transistors so that the resistance value contributed by the transistor is almost negligible and thus, the driver output impedance is primarily dominated by the series resistor, R_S. Moreover, large pre-driver is required to drive the large transistors in the output driver, which adversely impacts the area efficiency. Thus, to overcome these limitations, another type of SST driver of driver circuit is implemented by connecting the series resistor with a smaller resistance, R, in series with the pull-up and pull-down segment (**Figure 2a**) to improve the linearity of driver without area penalty [20]. This is because, the use of smaller resistance, R, can reduce transistor size and hence, further reduces silicon area occupied when compared to conventional SST driver shown in **Figure 1c**. However, the series resistor cannot be too small to the point that the driver starts to exhibit non-linear I–V characteristic. This is a tradeoff between driver linearity and driver silicon area. **Figure 3b** shows the increase of series resistance in pull-down segment (when input of SST driver is logic high) can

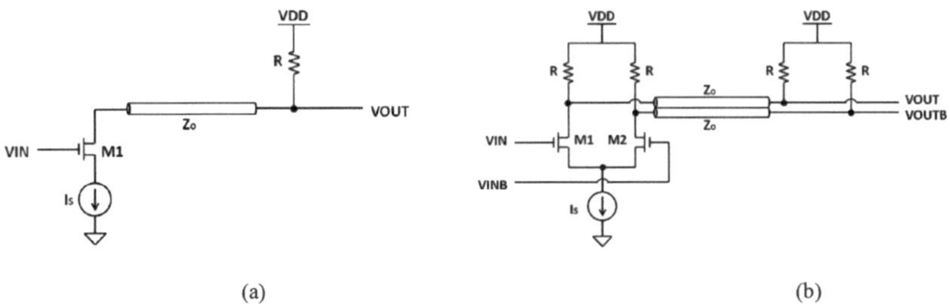

Figure 2.
(a) Saturated FET driver circuit with resistive load at receiving-end, and (b) differential current-mode driver with terminating load at transmitting-end and receiving-end [8].

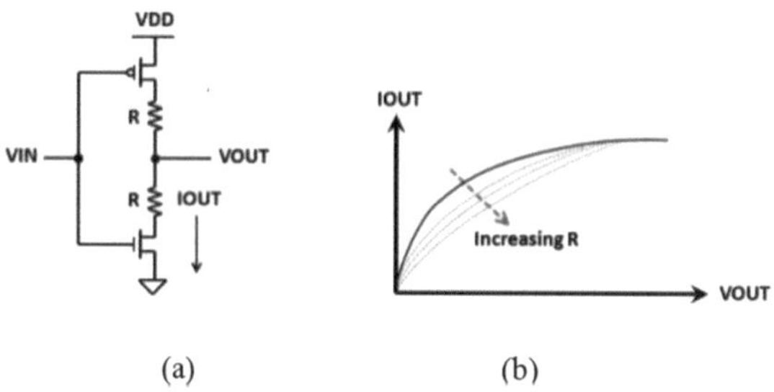

Figure 3.
(a) Architecture of SST driver that has series resistor R in pull-down and pull-up segment, and (b) I–V linearity is improved with larger series resistance R [20].

improve driver linearity. A suitable series resistance R must be selected to give a balance between driver linearity and silicon area.

2.1.2 Current-mode drivers

The simplest current-mode driver is a saturated FET driver shown in **Figure 2a**. From this figure, the impedance looking into the transmitting-end is huge because of the huge impedance from the constant current source implemented by the transistor in saturation region. This circuit is simple as it does not require any pull-up device at the transmitting-end and this feature makes it very suitable in low-cost applications [8]. Moreover, resistive load value, R, can be chosen as Z_O to match the transmission channel so as to minimize signal reflected back to transmitting-end. However, high-speed data transmission often employs both side termination schemes to minimize reflections. This makes the saturated FET driver unsuitable in high-speed application due to unmatched impedance at the transmitting end [26]. Differential current-mode driver can overcome the limitations of saturated FET driver [8]. The circuit of differential current-mode driver with resistive terminator, R, at transmitting-end and receiving-end is depicted in **Figure 2b**.

The trend in scaling down CMOS process technology dramatically deteriorates the channel-length modulation effect and this causes challenges in designing output driver circuits that employs current-mode transmission, such as the current- mode drivers. Channel length modulation causes symmetric error between current-mirror and current-mode driver, resulting to output swing that has large deviations from the desired value [24]. A non-optimal solution is to supply more current into output driver through the current-mirror to increase the output swing. However, this solution can cause larger power consumption [15]. Thus, current compensation (ICOMP) techniques have been used to overcome the symmetric errors between current mirror and current-mode driver to control output swing at desired value [24].

2.1.3 Configurable drivers

Fixed type of output drivers introduced in previous sub-sections might not be suitable in some circumstances as their advantages cannot be completely utilized to maximize the system performance. A fixed I/O driver is a compromise of various applications when it is used in multiple applications as the fixed features of that I/O driver cannot fit in various applications [9]. Besides, system upgrade that touches upon different signaling types and data rates has established design challenges on I/O drivers. A fixed type I/O driver is unable to support backward compatibility that come across with different signaling types and data rates [13]. For instance, I/O driver designed for Universal Serial Bus (USB) applications may be required to support different operating modes like low-speed, full-speed and high-speed mode that involve different signaling types and data rates [27, 28]. Therefore, configurable drivers are utilized in this situation. Different types of drivers can be integrated into one configurable driver so that the suitable driver type can be selected to fit the applications, as well as to support different signaling types and data rates. A multi-configurable HSTL (High-Speed Transceiver Logic)/LVCMOS (Low Voltage CMOS)/Open-Drain output driver circuit was proposed in Ref. [12] that consists of push-pull and open-drain transistor that are selectively enable or disable, depending on the desired mode of operation in a variety of system configurations. **Figure 4a** shows the circuit of multi-configurable output driver. Generally, this configurable I/O driver circuit can be switched into two main categories which are push-pull and open-drain mode through the selector transistors.

In push-pull mode, transistors P2 and N3 are deactivated, resulting in push-pull circuit as shown in **Figure 4b**. The push-pull circuit consists of pull-down transistor N2, pull-up transistor N1, and series terminator RT. Instead of using PMOS transistor, this circuit uses a NMOS transistor to pull-up the output voltage due to its fast switching speed. However, the transistor N1 can only charge the output voltage up to $VDD - V_T$, where V_T is the threshold voltage of PMOS. Thus, a PMOS transistor P1 is activated to further pulling-up the output voltage to VDD after a short delay introduced by inverter. In addition, the supply voltage VDD can be changed to different levels to support HSTL and LVCMOS interface when the driver is operated in push-pull mode. Also, in open-drain mode, transistors N1 and P1 are deactivated as shown in **Figure 4c**. Open-drain transistor N3 is used to transmit data while the transistors P2, N2, and terminator resistor RT are used to absorb the surge energy from the transmission channel when N3 is turned-off. However, the transition time of this multi-configurable I/O driver is not-well controlled and this causes the driver to be vulnerable to output oscillations. Full-swing nature can restrict the driver from operating toward higher data rates. Moreover, non-linear characteristics of output impedance makes this driver susceptible to reflection noise at high-speed.

Consequently, a dual-mode transmitter that is programmed and configured as full-swing voltage-mode transmitter for low speeds or a 400 mV reduced-swing current-mode for high speeds was proposed in Ref. [13]. **Figure 5a** shows the dual-mode transmitter that comprises of pre-driver and main driver. Pre-driver in **Figure 5a** decodes the mode control signals and generates necessary signals based on input data TXi for main driver. When the dual-mode transmitter manipulates in voltage transmission mode, transistor M1 and M3 switches in opposite phase with input, while M2

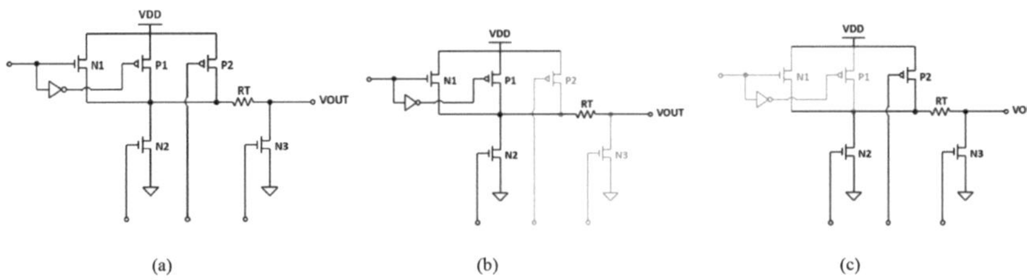

Figure 4.
(a) A multi-configurable HSTL/LVCMOS/open-drain output driver and multi-configurable driver in (b) push-pull mode, and (c) open-drain mode [12].

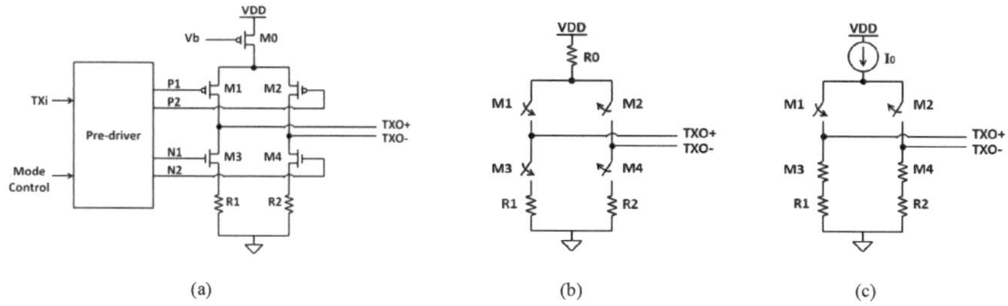

Figure 5.
(a) Dual-mode transmitter circuit; and dual-mode transmitter operates in (b) voltage-mode, and (c) current-mode [13].

and M4 switches in phase with the input (**Figure 5b**). Transistor M0 operates in the linear region which serves as the pull-up terminate resistor, R0, whereas resistor R1 and R2 are pull-down terminators, as depicted in **Figure 5b**.

In the current transmission mode shown in **Figure 5c**, transistors M1 are M2 are switched differentially, while M3 and M4 always operate in linear region which can combine with R1 and R2 to serve as terminate resistors. Transistor M0 operates in saturation mode as a current source biased by Vb. However, this transmitter suffers from the transistor channel-length modulation effect that causes large symmetric error between current mirror and current-mode driver. Moreover, continuous scaling down of CMOS technology has dramatically increases the channel-length modulation effect and this can cause the transmitter output swing to have large deviations from the desired value and result in wrong logic sampled at receiving-end.

2.2 Techniques to improve driver circuit performance

The conventional ways to improve I/O performance include using differential signaling to reduce common-mode noise, controlling output slew rate to reduce switching noise, using on-die termination to absorb signal reflections, controlling output impedance to minimize signal reflections, controlling output swing of current-mode driver to minimize output swing deviations, as well as de-emphasizing data to overcome inter-symbol interference (ISI).

Differential signaling improves the signal-to-noise ratio (SNR) of a driver circuit as it causes the swing of differential signal to be doubled when compared to the swing of a single-ended signal. Moreover, majority of common-mode noise can be rejected through the use of differential signaling in data transmission which further improves the performance of the driver circuit [8, 13].

Simultaneous switching noise (SSN) and ground bounce noise are important considerations in the contemporary output driver circuit design [19, 28]. In general, the purpose of controlling slew rate of output driver is to have a controllable switching rise and fall times from the I/O driver [20]. Some of the several techniques to control the output slew rate include the use of passive delays after pre-driver stages; and the time-split technique. The former slows down the transition time of the I/O drivers and the output slew rate to reduce SSN [8]. However, this techniques may lead to large delay and results to excessive ISI because the pre-driver output will not reach steady state by the end of bit cell and may result to worst jitter [20]. Hence, this concept is not feasible for I/O design at the present-day. The latter is based on the principle of splitting the pull-down and pull-up transistors into parallel segments and sequentially activating each segment in a proper delay, which is considered a very effective technique and is suitable for high speed CMOS drivers to reduce both SSN and reflection noise at the same time [13].

In some high-performance signaling systems, resistive terminations are needed at both transmitter and receiver to absorb traveling waves to prevent unwanted reflections [29]. In these systems, the receiver termination absorbs the forward-traveling signal transmitted by transmitter, while transmitter termination absorbs reflections from transmission channel impedance discontinuities to prevent backward-traveling reflections [8]. In the communication between chips, the receiver termination, RTERM, can be achieved by placing discrete resistors on package or board and this can be called off-chip termination. In addition, instead of placing the necessary termination on the package, the advancement in fabrication process has enabled the termination to be located on the chips or dies and this technology is called on-die termination (ODT), which allows simpler system design and cost effective as the employment of ODT can reduce the number of resistive termination elements and complex wiring on the board [8]. The ODT can be easily implemented by using transistors or combinations of transistors and resistors from CMOS fabrication processes. However, these transistors and resistors are PVT (Process, Voltage, and Temperature) dependent and they must be compensated by on-chip resistance calibration technique.

On-chip resistive termination can be used to absorb traveling waves and prevent unwanted reflections at transmitters, receivers or both sides in high performance systems. Hence it is critical to have the impedance terminator always close to the impedance of transmission channel over all possible temperatures, voltages and processes [30]. As discussed earlier, the SST driver is formed by a transistor connected in series with a resistor. During fabrication, as they go through the myriad stages like photolithography, ion implantation, doping, metallization, and oxides, all contributes to various aspects of the final circuit [8]. For example, an oxide thickness of the devices may result to different I–V characteristics of transistors. Thus, if such device with inconsistent I–V characteristic is used as driver element, large variations in driver strengths will occur and impact the system performance due to un-matched impedance value of driver with the transmission channel [31]. Therefore, it is necessary to calibrate the on-chip terminators by either analog or digital means over all possible ranges of PVT [23, 30]. While the advantages of analog resistance calibration is the fact that the resistance value is continuously adjusted according to PVT variations and it does not generate noise during the calibration process [23], the advantages of the digital calibration scheme are the on-chip DC power reduction, owing to the employment of digital gates [19]. However, the analog technique in Ref. [23] has high DC power dissipation and the impedance of several on-chip drivers cannot be adjusted independently [30].

The output swing of current-mode driver is required to be controlled across PVT variations to provide sufficient eye opening at far-end without consuming excessive power. Thus, the tail current source used to generate the driver output swing needs to be PVT-compensated against the desired reference swing [24]. Hence, the Current Compensation (ICOMP) block is proposed by Ref. [24] to calibrate driver output swing across PVT variations.

Equalization is a circuit technique to compensate for the loss of transmission channel at high speed [29, 32]. Equalization circuits can be integrated with transmitter circuits or receiver circuits. Transmitter equalization is commonly done in high performance systems because transmitter equalization permits the use of simpler far-end receiver as compared to receiver equalization that needs a high resolution ADC (analog-to-digital converter) that creates design challenges [8]. Some transmitter equalization circuits include pre-emphasis and de-emphasis techniques [33]. Pre-emphasis technique emphasizes the high-frequency components of the transmitted signal. The pre-emphasis transmitter increases the amplitude of high-frequency components or transition pulses as compared to the amplitude of high-frequency components in the original signal swing. However, the increase of amplitude of high-frequency components can cause EMI (Electromagnetic Interference) and more severe current spikes [33]. Therefore, de-emphasis technique is developed to minimize EMI and current spikes problems by only de-emphasizing the low-frequency components of a digital signal [30–34]. More information about de-emphasis types (differential current-mode driver and differential voltage-mode driver) are presented in Ref. [15].

3. Design methodology and circuit implementation

This chapter presents the design and implementation of a configurable I/O driver as a standard cell that can achieve optimum performance across various ASIC applications. The configurable I/O driver can be configured in different modes supporting a variety of intended ASIC applications. The proposed driver is integrated with impedance, output slew rate, de-emphasis and output swing control schemes to address reflection noise, minimize switching noise, compensate ISI noise and control output swing at desired level, respectively.

3.1 Circuit simulation model

The proposed driver performance is verified through simulates with non-ideal characteristics that are commonly used in modern IC packages. These non-ideal characteristics are the package and substrate parasitic that cause I/O ringing [19] and the transmission channel losses at high frequency [32]. The package parasitic and transmission channel are modeled based on [29, 35–37].

3.2 Performance evaluation methodology

By including the non-ideal characteristic like package parasitic and transmission channel losses into simulation, the performance of the proposed configurable I/O driver is evaluated using eye diagram with a series of random data generated from 27 – 1 PRBS (Pseudo Random Binary Sequencer) [38]. To create eye diagram, the time-domain signal waveform is slicing into segments and overlaying them. Therefore, in this work, the eye jitter must be smaller than 0.30 UI to provide sufficient time to fulfill setup and hold constraint of far-end receiver. Moreover, the maximum height of differential eye must be larger than 200 mV to ensure that the voltage levels meet input high and low limit requirements of the receiver [9]. It is reasonable to restrict the differential eye height and jitter specification as this allows the use of a relatively simpler and more power efficient receiver at far-end.

3.3 System architecture and performance requirements

The configurable I/O driver design in this research is established according to the specifications stated in **Table 1** [24, 39].

Specifications	Value
Supply voltage (VCC)	1.0 V ± 5%
Data rates	0.4–8.0 Gbps
Output slew rates	2–4 V / ns
ODT impedance	(50, 75 and 150 Ω) ± 5%
ODT impedance linearity error	± 15% maximum
Output driver impedance	50 Ω ± 5%
Output driver impedance linearity error	± 15% maximum
Output swing (single-ended in current-mode driver)	(400 and 500 mV) ± 5%
Receive eye jitter (peak-to-peak)	0.30 UI maximum
Receive eye height (at middle of differential eye)	200 mV minimum

Table 1.
Specifications of the proposed configurable I/O driver.

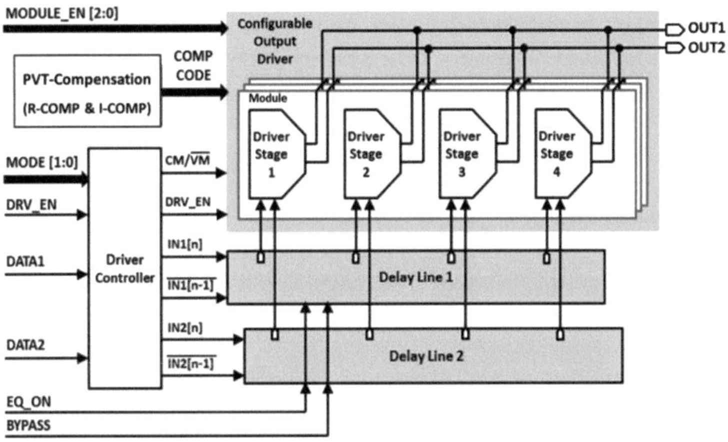

Figure 6.
Overall system block diagram.

The overall architecture of proposed system consists of a configurable I/O driver, a driver controller, two delay lines (Delay Line 1 and Delay Line 2), and a PVT-compensation block that contains R-COMP and I-COMP, as depicted in **Figure 6**.

The proposed I/O driver in **Figure 6** can be configured into different driver modes to support different signaling and achieve optimum performance across different processor-intended applications. In the driving mode, the proposed I/O driver can be configured into single-ended/differential voltage-mode driver (push-pull driver) or differential current-mode driver. In the receiving mode, the proposed I/O driver can be configured as pull-up or pull-down ODT to terminate any reflections from far-end, thereby improving signal integrity for near-end receiver. The proposed I/O driver consists of three driver modules and is designed, based on a modular concept. The driver is subdivided into smaller modules and each module is further decomposed into four driver stages that can be designed independently. Modular design can ease the design process and allow different impedance values to be configured from the I/O driver. Moreover, the pull-up and pull-down segment in each driver stage are

PVT-compensated at 600 Ω. This gives an equivalent output impedance of (600 Ω/4 stages) or 150 Ω when all the four stages in a module are activated. Thus, by enabling different number of driver modules in parallel, through the MODULE_EN [2:0] signal, multiple impedance values of 50, 75 or 150 Ω can be obtained. Also, the driver controller in **Figure 6** is used to process and prepare the input/output driver. Input and output can be two single-ended data or differential data, depending on mode control signal MODE[1:0].

The output slew rate control scheme is applied in the proposed system in **Figure 6**. In this control scheme, two delay lines, namely, Delay Line 1 and Delay Line 2, are used to control the output slew rate to reduce switching noise. In addition, when the output driver is operating with high data rate over long transmission channel, the ISI noise starts dominating switching noise and de-emphasis control scheme can be introduced to reduce ISI-induced jitter; thereby reducing sampling errors and improving signal integrity over the transmissions.

The PVT-compensation block in **Figure 6** consists of impedance compensation (R-COMP) and current compensation (I-COMP) block implemented based on on-chip impedance and output swing control scheme, respectively. The R-COMP is used to calibrate on-chip impedance within ±5% variation as shown in **Table 1**, across PVT variations to minimize reflection noise. When the proposed I/O driver is configured as current-mode driver, I-COMP block can be used to calibrate output swing within ±5% variation from targeted swing across PVT variations. All compensation blocks are global blocks and their digital codes can be distributed within the chip.

3.4 Design architecture

This section presents the design details of each block in the overall system block diagram (**Figure 6**), including configurable I/O driver, driver controller, delay lines and compensation block.

3.4.1 Configurable I/O driver

As described in Section 3.3, the configurable I/O driver is designed based on modular design concept such that it can be divided into smaller driver stages that can be created independently. **Figure 7** shows the architecture in a driver stage that consists of pre-driver, mode-controller and main driver. The pre-driver provides suitable VP1, VP2, VN1, VN2 and VSW signal based on the driver control signals generated by driver controller. Main driver consists of two half drivers with a switch M7 between them. Series resistor R1 and R2 are sized to 85% of overall impedance to reduce the linearity error of pull-down and pull-up segment below 15% [22]. This is essential because the I/O driver should act like a self-terminator that can eliminate any reflections in high speed communications [26]. Thus, the impedance of both pull-up and pull-down segment are needed to exhibit linear characteristics (**Figure 3**) and match transmission channel impedance to terminate any reflections. Also, the operations when the I/O driver is configured into different driver types will be discussed in the following sub-sections.

3.4.1.1 Single-ended and differential voltage-mode driver

Main driver in **Figure 7** can be configured as two single-ended voltage-mode drivers implemented using the push-pull topology. The operation of the two single-

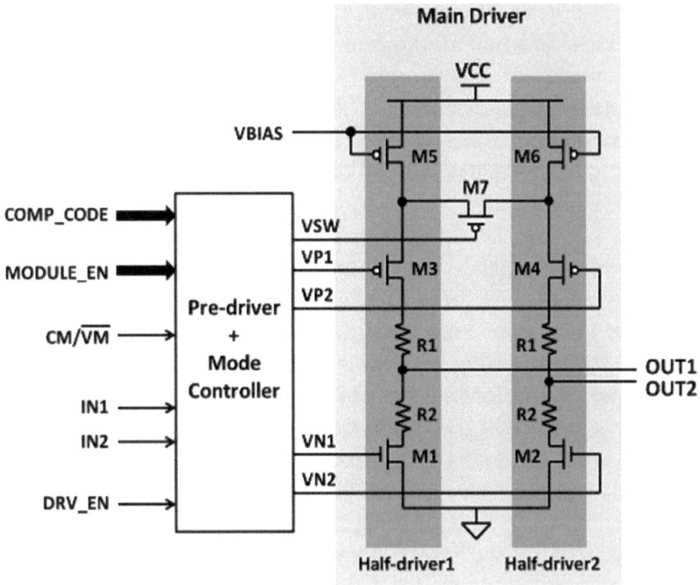

Figure 7.
Pre-driver and main driver architecture in a driver stage.

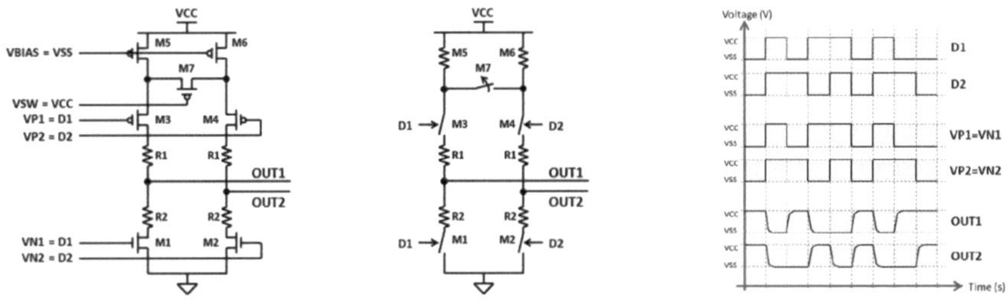

Figure 8.
Single-ended push-pull driver configured from main driver: (a) transistor implementation, (b) macro-model, and (c) operation in timing diagram.

ended push-pull drivers is shown in **Figure 8a**, where D1 and D2 are two independent inputs generated by a pre-driver. In push-pull operation, input transistor M1 and M3 are switched in phase with D1, while input transistor M2 and M4 are switched in phase with D2. Transistor M5 and M6 operate in linear region as the biasing voltage VBIAS is set to VSS. Switch M7 is opened by setting VSW = VCC to ensure proper push-pull operations for the two single-ended push-pull drivers. **Figure 8b** shows the macro-model of two single-ended push-pull drivers, while **Figure 8c** is a timing diagram to demonstrate the operation of single-ended push-pull driver. However, the full swing nature of single-ended push-pull driver limits the data rates [13]. Thus, this type of driver is only suitable for some low-speed and low-cost applications where terminators at receiving ends can be removed to reduce system costs.

However, low swing and differential signaling are keys to improve data rates. Low signal swing can reduce current spikes, transition time, while differential signaling can reject common-mode noise along the transmission. **Figure 9a** shows the transistor

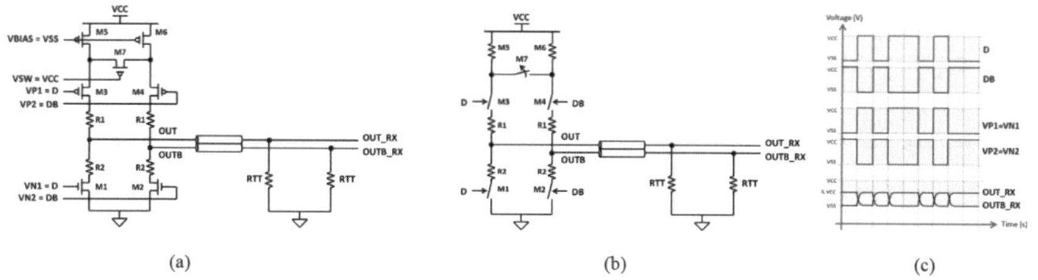

Figure 9.
Differential push-pull driver: (a) transistor implementation with far-end termination to ground, (b) macro-model, and (c) operation in timing diagram.

implementation of differential push-pull driver. The operation is same as that in the single-ended push-pull driver, except that the two inputs now become differential input (D and DB) generated from the pre-driver. Differential push-pull driver can support two termination schemes at far end, namely, low common mode and high common mode terminations and hence, the signal swing is half of the supply rails and this allows the output driver to operate at higher data rates. The macro-model of differential push-pull driver is shown in **Figure 9b**. In addition, the operation of differential push-pull driver is also demonstrated in **Figure 9c**.

3.4.1.2 Differential current-mode driver

Main driver in **Figure 7** can be configured as a differential current-mode driver. **Figure 10a** shows the transistor implementation of differential current-mode driver.

From **Figure 10a**, pull-down segments serve as transmit terminators with transistor M1 and M2 biased at VCC by the pre-driver. Transistor M5 and M6 are biased to be current sources through current-mirrors while transistor M7 acts as a closed switch. Input transistor M3 and M4 function as switches to alter the direction of current flow from current sources based on the differential inputs (D and DB). Moreover, the operation of current-mode driver can be explained using macro-model and timing diagram in **Figure 10b** and **c**, respectively. When D is low, transistor M3 is turned-on while M4 is turned-off. This allows a total current of $2 \times I_B$ flow through M3 to both the pull-down segment and far-end RTT with each carry current amount of I_B to raise the node, OUT, to the desired voltage swing level, V_{SWING}. To maintain output high level at V_{SWING}, current I_B can be tuned and calibrated in each driver stage across different PVT corners. By calibrating I_B in each driver stage, one interesting feature is that the output high level can still be maintained at V_{SWING}, when all the three driver

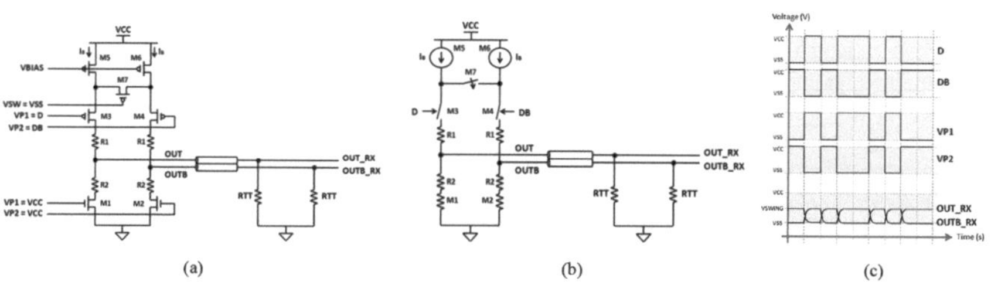

Figure 10.
Differential current-mode driver with far-end termination resistors: Transistor implementation, (b) macro-model, and (c) operation in timing diagram.

modules are enabled. This is because, every driver stage in the driver modules has identical operating points. Thus, instead of three driver modules, it is sufficient to use only one driver stage in I-COMP and this can significantly reduce power consumption and silicon area footprint.

3.4.1.3 Receive on-die termination (ODT)

When the proposed configurable I/O driver is operating in receiving mode, it can be configured as receive ODT to absorb any reflected signals to improve the signal integrity for near-end receiver. This can be achieved by enabling different number of 150 Ω driver module in parallel through the MODULE_EN[2:0] signal in **Figure 6**, thereby providing multiple impedance values of 50, 75 or 150 Ω. Thus, the main driver in **Figure 7** should be designed such that it can be configured as near-end receive ODT to power supply or ground. **Figure 11a** shows the transistor implementation of near-end receive ODT.

In addition, the marco-model of receive ODT and its operation in time domain is shown in **Figure 11b** and **c**, respectively. The advantage of this receive termination-merged driver structure is that it helps in reducing the output pin capacitance, when compared to conventional implementations which have separate structure for the transmit drivers and receive termination [40].

3.4.1.4 Pre-driver and mode control circuit

The main driver in **Figure 7** consists of half-driver1 and half-driver2 circuit. Two pre-drivers are required to generate the corresponding inputs for the half-drivers (VP1 and VN1 for half-driver1, VP2 and VN2 for half-driver2). Thus, **Figure 12a** shows the circuit of pre-driver for a half-driver. By taking input data (PU and PD), driver enable (DRV_EN and DRV_ENB), driver module enable (MODULE_EN and MODULE_ENB) and compensation code (COMP_PU and COMP_PD), the logic gate NAND1 and NOR1 in pre-driver can control the main driver to pull-down/pull-up the output or in tri-state condition. In addition, the mode controller in the pre-driver can generate suitable input data PU and PD based control signal CM/VM for different operation modes i.e. voltage-mode (push-pull driver), current-mode driver or termination, as shown in **Figure 12b**. The operations of mode controller for different drive types are also summarized in **Table 2**.

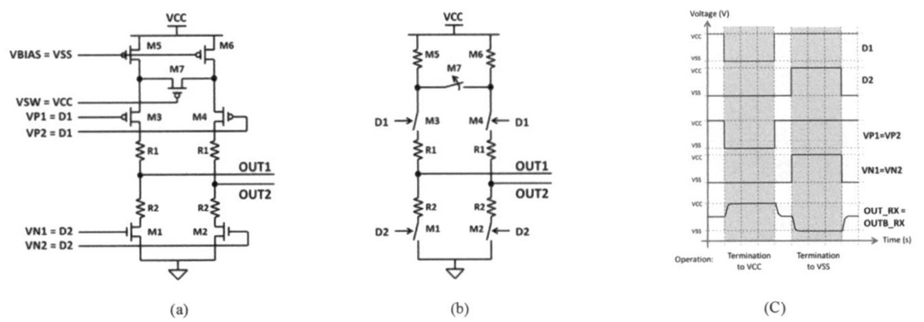

Figure 11.
Near-end receive ODT configured from main driver: Transistor implementation, (b) macro-model, and (c) operation in timing diagram.

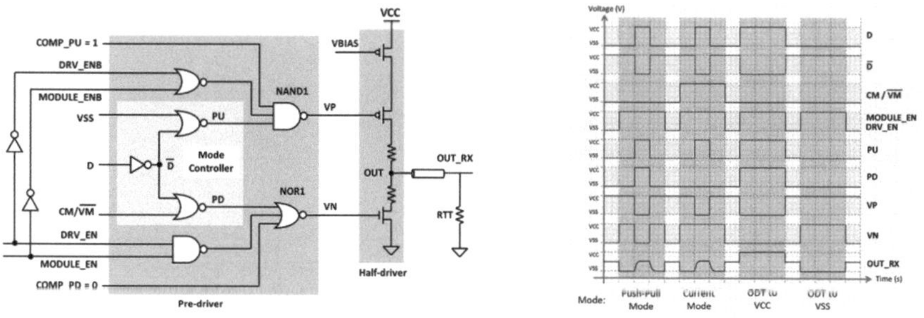

(a) (b)

Figure 12.
Pre-driver and mode-controller for a half-driver circuit in ***Figure 12****: (a) logic implementation, and (b) operation sequences in time domain.*

Operating Mode	Input	Output	
	CM/VM	PU	PD
Push-pull	0	D	D
Current-mode	1	D	VSS (pull-down segment always ON)
ODT to VCC*	0	D = VCC	D = VCC
ODT to VSS*	0	D = VSS	D = VSS

**Note: The operation of receive ODT is the same as push-pull driver, except input D is set to power rails.*

Table 2.
Driver operation mode and output of mode controller circuit.

3.4.2 Driver controller

The functional diagram of driver controller is shown in **Figure 13**.

The main function of driver controller is to decode MODE[1:0] to configure the proposed I/O driver into desired operating driver types. In **Figure 13**, the driver controller consists of de-emphasis block, command decoder block and data controller block. De-emphasis block is employed to generate inverted one bit shifted data $\overline{IN1[n-1]}$ and $\overline{IN2[n-1]}$ based on IN1[n] and IN2[n] for the de-emphasis driver when the de-emphasis control scheme is activated. Command decoder block is used to decode the MODE[1:0] into three control signals, namely CM/$\overline{VM}$, OCD/$\overline{ODT}$ and DIFF/$\overline{SE}$. The descriptions of these control bits are shown in **Table 3**.

Also, the logic implementation of the data controller block is also shown in **Figure 14**. The function of data controller is to generate ordinary data IN1[n] and IN2[n] based on control signal OCD/$\overline{ODT}$, DIFF/$\overline{SE}$ and ODT_VCC/$\overline{ODT\_VSS}$ for the proposed I/O driver. When OCD/ODT is asserted high, the proposed I/O driver is in driving mode and IN1[n] and IN2[n] can be generated as differential data or two single-ended data based on DIFF/$\overline{SE}$ bit. When OCD/$\overline{ODT}$ is asserted low, the proposed I/O driver is in receiving mode and both the IN1[n] and IN2[n] can be generated as VCC or VSS for receive ODT to power supply or ground based on ODT_VCC/$\overline{ODT\_VSS}$ bit. The operations of data controller are demonstrated as a flow chart in **Figure 14**.

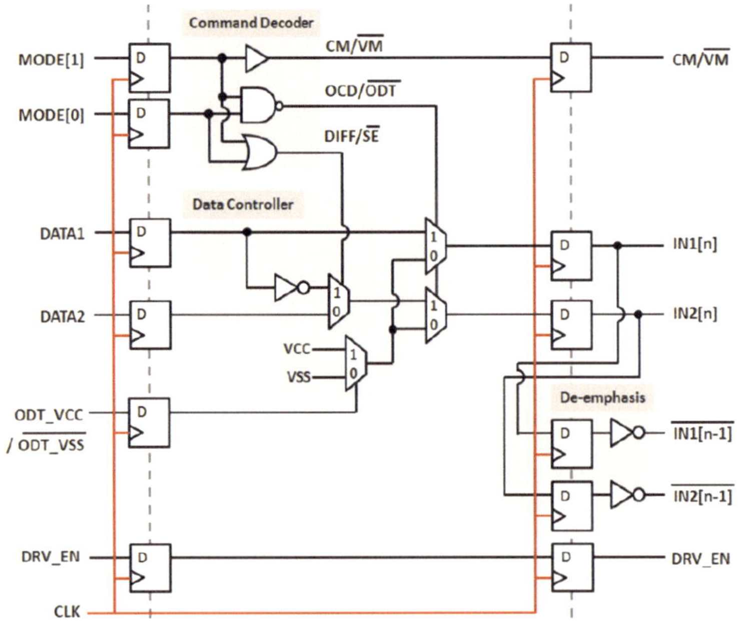

Figure 13.
Block diagram of driver controller in logic implementation.

Control bit	Description
CM/$\overline{\text{VM}}$	Asserted high if current-mode driver is configured. Otherwise, voltage-mode driver is configured.
OCD/$\overline{\text{ODT}}$	Asserted high if proposed design is configured as off-chip driver. Otherwise, the proposed design is configured as receive on-die terminator.
DIFF/$\overline{\text{SE}}$	Asserted high if differential signaling is configured. Otherwise, single-ended signaling is configured.

Table 3.
Description of control bits generated by command decoder.

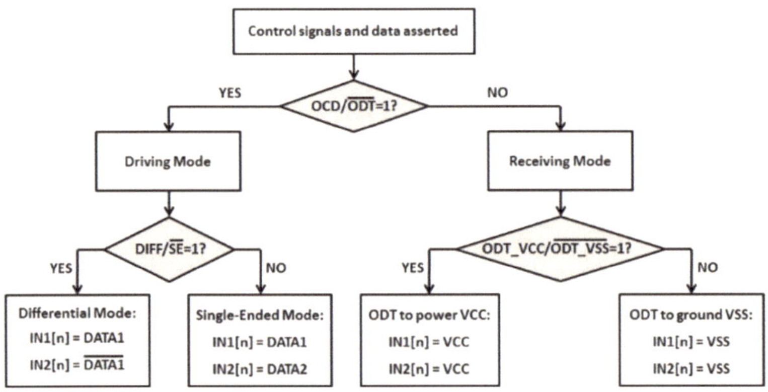

Figure 14.
*Operations of data controller based on control bits in **Table** 2.*

3.4.3 PVT-compensation block

On-chip impedances and output swing (in current-mode driver) are subjected to PVT variations and are required to be calibrated to sustain the driver performance. In this work, the main driver in the driver stage (**Figure 7**) is designed with multi-leg structure where the pull-up and pull-down segment consists of one static leg and four binary-weighted dynamic legs that can be turned-on or turned-off for impedance and output swing calibration, as depicted in **Figure 15a** and **b**. Impedance control and output swing control scheme can be applied by adjusting the dynamic legs of the multi-leg structure in the PVT-compensation block shown in **Figure 6** to calibrate driver impedance and output swing, respectively.

3.4.3.1 Impedance control scheme

Impedance control scheme is implemented by R-COMP block in the PVT-compensation block to calibrate the driver output impedance. Digital compensation is chosen due to the merit of DC power reduction with digital logics in R-COMP. In the R-COMP block, the driver impedance of a single driver stage (see **Figure 7**) is compensated at 600 Ω ± 5% by using a dummy pull-up and pull-down segment shown in **Figure 16a** and **b** respectively. **Figure 16a** shows the architecture of compensation block for pull-up impedance. The output net VOUTP from pull-up segment is connected to an external reference resistor REXT with 600 Ω ± 1%. Each pull-up segment has one static leg and four-bit binary-weighted dynamic legs in parallel. Static leg is always activated to provide impedance that are close to 600 Ω to fasten

Figure 15.
Main driver in **Figure 7** *designed with multi-leg structure for (a) pull-up segment, and (b) pull-down segment.*

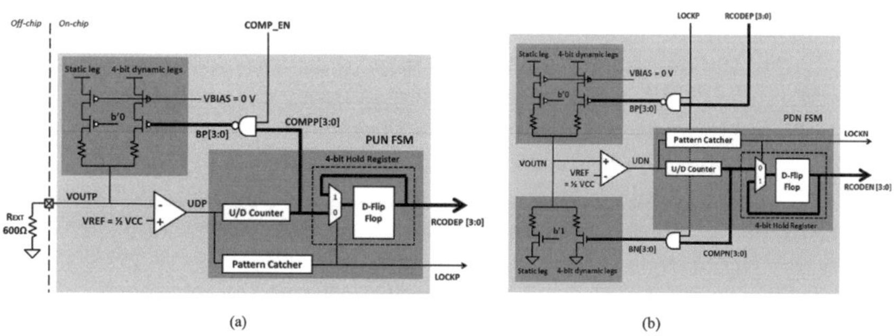

Figure 16.
Impedance compensation block (R-COMP) for (a) pull-up segment, and (b) pull-down segment.

calibration process, while dynamic legs are adjusted by PUN FSM (Pull-Up Network Finite State Machine) until the total impedance match to 600 Ω. When VOUTP is lower than VREF, this implies pull-up impedance is larger than reference resistor. This delta voltage will cause comparator outputs UDP with logic high for up- down counter (U/D Counter) to count up the COMPP[3:0] so as to increase the dynamic legs in the pull-up segment to rise the VOUTP voltage until it is slightly higher than VREF. Then, the comparator will output UDP with logic low for U/D Counter to count down COMPP[3:0] to decrease the dynamic legs by one and hence, reducing VOUTP. This process will continue and the VOUTP will be alternating about VREF as UDP always toggle due to the quantization error between pull-up impedance and reference resistance. This implies that the COMPP[3:0] will be increased or decreased by one and this toggling code cannot be delivered to I/O driver. Thus, a pattern catcher is used to catch the up-down sequence of the UDP signal. When VOUTP is alternating about VREF and UDP is toggling, the pattern catcher will output a lock signal (LOCKP) to the 4-bit hold-register to make the RCODEP[3:0] = COMPP[3:0] at the output of register so that the compensation code is held constant for the I/O drivers.

Figure 16b shows the architecture of compensation block for pull-down impedance. Once the compensation for the pull-up segment is done, the LOCKP signal will be generated by PUN FSM to kick-off the compensation process for pull-down segment. In addition, the RCODEP[3:0] from PUN FSM will be delivered to the pull-down segment in **Figure 16b** since this pull-up segment (compensated at 600 Ω) is now becoming reference resistor for the pull-down segment. When the VOUTN is higher than VREF which implies pull-down impedance is larger than 600 Ω, the comparator will output UDN with logic high for U/D Counter to count up the COMPN [3:0] to increase the number of dynamic legs in pull-down segment so as to reduce the VOUTN voltage. This searching process will continue until VOUTN is alternating about VREF and UDN is toggling, the pattern catcher will output a lock signal (LOCKN) to 4-bit hold-register to hold the output at RCODEN[3:0] = COMPN[3:0]. Hence, the pull-up and pull-down calibration codes, RCODEP[3:0] and RCODEN [3:0], respectively, are ready to be distributed to all the I/O drivers in the chip.

3.4.3.2 Output swing control scheme

The proposed configurable I/O driver in current-mode can support output swing V_{SWING} of 400 and 500 mV. However, actual output swing can vary and deviate from the desired swing level, V_{SWING}, which can cause sampling error at the far-end receiver. This is mainly due to the variation of the tail current across PVT variations and also the symmetric error between biasing circuit and tail current sources. Output swing control scheme is implemented by the I-COMP block in the PVT-compensation block to calibrate the tail current so that the driver output swing can be maintained at $V_{SWING} \pm 5\%$. In I-COMP, a dummy driver stage is used as a compensation with one of the input transistor, which is turned-off, to allow current from both the tail current sources flow to another input transistor to the two pull-down terminators, as shown in **Figure 17**. Pull-down terminators are calibrated by pull-down code RCODEN[3:0] from R-COMP block and one of them is a dummy receiver terminator to emulate the situation where the driver is operating with double terminations. VOUT is compared to the reference voltage VREF (can be 400 or 500 mV) and the dynamic legs inside the pull-up segment will be adjusted by ICOMP FSM to increase or decrease the current amount that flow to the pull-down terminators and thus, controlling output swing at VREF. The compensation process is similar to impedance compensation in R-

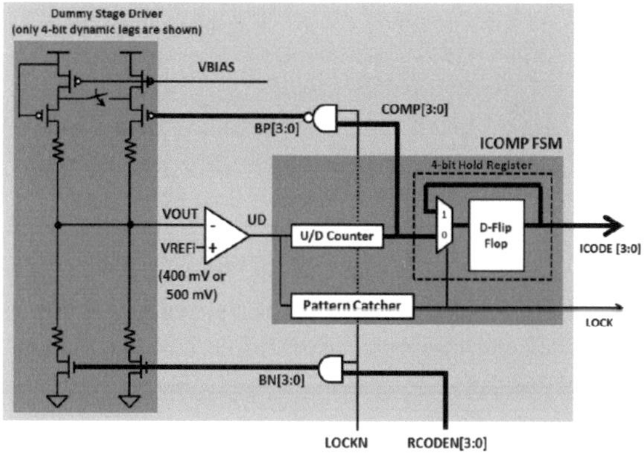

Figure 17.
Current compensation I-COMP block.

COMP block. When compensation for pull-down segment is done, LOCKN signal will be generated by PDN FSM to kick-off the output swing compensation process for the pull-up segment. If the VOUT is lower than VREF, which implies that there is insufficient current from tail transistors in pull-up segment, then the comparator will output UD with logic high for U/D Counter to count up the COMP[3:0]. This increases the number of dynamic legs in the pull-down segment so as to increase the tail current, as well as the VOUT voltage. This searching process will continue until VOUT is alternating about VREF and UD is toggling, then the pattern catcher will output a lock signal (LOCK) to 4-bit hold-register to hold the calibration code at ICODE [3:0] = COMP[3:0].

In this work, the design technique of I-COMP is similar with R-COMP (see **Figure 16**). The bias voltage VBIAS is constant from global current-mirror circuit and the dynamic legs are adjusted to calibrate the amount of current from tail current source, thereby calibrating output swing at the desired level. Moreover, ICODE[3:0] is a digital code that is insensitive to noises and this ensures the calibration information will not be impacted during delivery to other current-mode drivers for output swing calibration. This is different from technique proposed by Ref. [24] where the bias voltage ISOURCE from IDAC is delivered to local bias generators in all current-mode drivers. This noise-sensitive analog bias voltage, ISOURCE, in combination with different PVT variations in each local bias generator might result in large deviation on the output swing of each current-mode driver. Thus, the advantage of I-COMP in this work is the use of digital calibration technique to avoid the digital calibration codes from being impacted by noises and PVT variations.

3.4.4 Delay line architecture

3.4.4.1 Slew rate control scheme

Output slew rate control scheme is applied in the proposed I/O driver. Digital time domain splitting technique is used since the proposed I/O driver requires a stable control of slew-rate and output impedance. **Figure 18a** shows a diagram of a driver module designed using time split technique (simplified from **Figure 6**). The output driver in each driver module is segmented into four driver stages and the switching time of these driver stages is controlled through a delay line that consists of delay cells

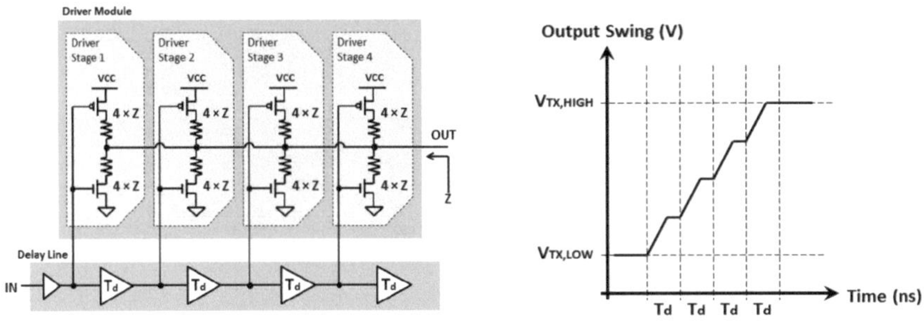

(a) (b)

Figure 18.
Time split technique: (a) simplified diagram of one driver module with delay line, and (b) controlled output slew rate.

with Td delay interval. By activating each driver stage in sequence, output transition time can be controlled. The pull-up and pull-down impedance in each driver stage is controlled at four times larger than the desired module impedance Z (150 Ω), hence, the effective output impedance is always constant at 50 Ω when three driver modules are switching together. The output slope depends on interval Td as shown in **Figure 18b**. Moreover, the driving single-ended output swing can be 500 mV in differential voltage-mode driver and 400 or 500 mV in the current-mode driver. More detailed information about a delay line can be found in Ref. [41].

As the delay across each delay cell is controlled at 40 ps, the output transition time can be controlled approximately at 160 ps, as depicted in **Figure 18b**. However, this limits the output driver from operating beyond 4 Gbps because the controlled output rise and fall transition time is occupying large fraction of the one bit-pulse duration, which can cause sampling errors at the far-end receiver. Thus, slew rate control scheme is only applied to applications with data rates below 4 Gbps. For data rate higher than 4 Gbps, the delay cells should be bypassed so that the output transition is no longer restricted by the delay intervals, although this could not help in reducing switching noise. However, as the output driver operates at higher data rates over long transmission channel, ISI noise start dominating switching noise and de-emphasis control scheme should be activated to reduce ISI-induced jitter and thus, improving signal integrity. **Figure 19** shows the architecture of delay line (Delay Line 1 and Delay Line 2 in **Figure 6**) where multiplexers act as bypass circuit. A multiplexer is added at the output of each delay cell and controlled by signal BYPASS. When BYPASS is low, the multiplexers select output of delay cells for driver stages in the configurable I/O

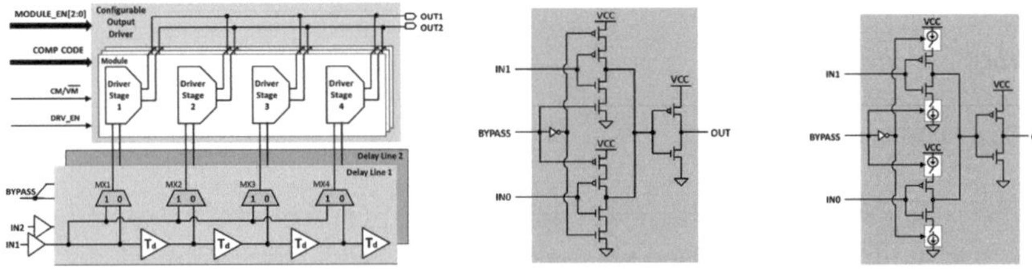

(a) (b) (c)

Figure 19.
*(a) Structure of delay line in **Figure 6** with delay bypass circuit, (b) multiplexer circuit using current-starved inverter [39], and (c) macro-model of multiplexer.*

drivers. If BYPASS is high, delay cells are bypassed and the multiplexers select IN1 and IN2 for all the driver stages.

Moreover, **Figure 19a** shows the circuit implementation of multiplexer that is designed based on the topology of current-starved inverter [42] while **Figure 19b** shows the macro-model of multiplexer. The merit of this topology is to allow faster switching at higher data rates since the current sources in current-starved inverter can quickly charge or discharge voltage in each internal node.

3.4.4.2 De-emphasis control scheme

When the I/O driver operates at higher data rates over long transmission channel, ISI becomes significant and de-emphasis control scheme should be activated to compensate the ISI impacts on signal eye jitter and eye height. **Figure 20** shows the generic idea of integrating de-emphasis scheme into the proposed configurable I/O driver. Similar to the basic concept of de-emphasis, the first three driver stages in a driver module can be connected as main tap driver while the last driver stage is connected as de-emphasis tap driver.

Moreover, in order to drive the main tap driver and de-emphasis tap driver with the corresponding ordinary data (IN1[n] and IN2[n]) and inverted one bit shifted data ($\overline{\text{IN1}[n-1]}$ and $\overline{\text{IN2}[n-1]}$), the delay line structure in **Figure 19a** must be modified accordingly. **Figure 21** shows the structure of delay line modified from **Figure 19a** to support de-emphasis control scheme. A multiplexer, MXEQ, with selector signal, EQ_ON, is added at the output path of multiplexer MX4 for the de-emphasis tap driver. When de-emphasis is required, BYPASS is set to high so that multiplexer MX1, MX2 and MX3 outputs ordinary data IN1[n] and IN2[n] to the main tap drivers, while EQ_ON is also set to high for multiplexer MXEQ to outputs inverted one bit shifted data $\overline{\text{IN1}[n-1]}$ and $\overline{\text{IN2}[n-1]}$ to the de-emphasis tap driver. The de-emphasis ratio of the proposed configurable I/O driver can be calculated as $-20\log[(3–1)/(3+1)] \approx 6$ dB [43]. In addition, the advantage of this de-emphasis driver is that it has a constant de-emphasis of 6 dB that is independent of output impedance. This is because, the de-emphasis ratio is only dependent on the number of instances in the main tap and post-cursor tap driver, while the output impedance is constant by calibrating the driver stage only. This gives a constant de-emphasis across PVT variations as compared to pre-emphasis transmitter proposed by Ref. [43], where the number of instances in the main tap and post-cursor tap driver are not always constant due to the need of output impedance tuning.

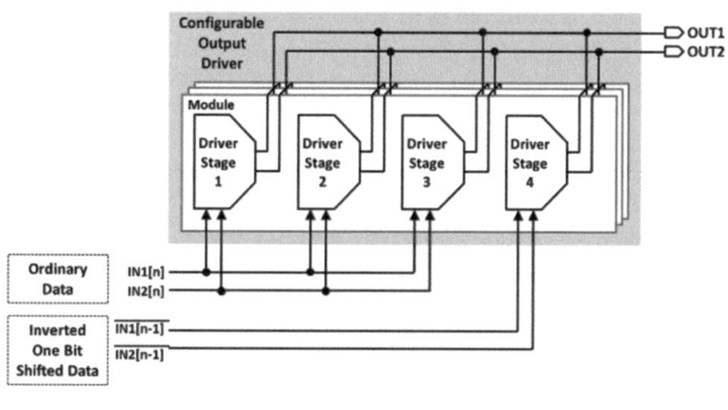

Figure 20.
Generic idea of integrating de-emphasis scheme into the proposed I/O driver.

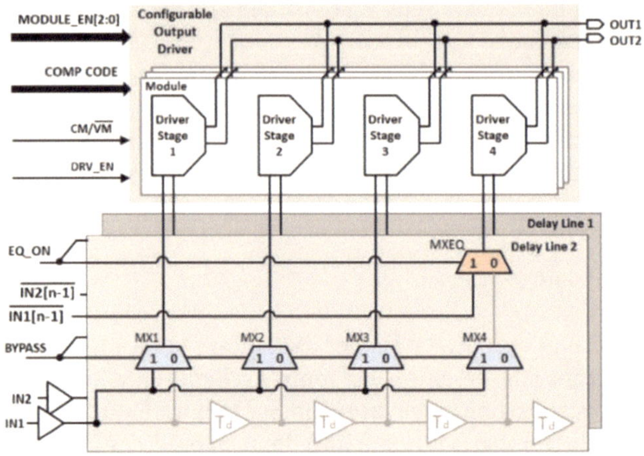

Figure 21.
Structure of delay line modified from ***Figure 19a*** *to support de-emphasis control scheme.*

3.5 Layout implementation

The research development moves forward to layout implementation once the proposed I/O driver is successfully designed and simulated. Post-layout simulations are then performed to ensure the layout parasitic do not significantly affect the I/O driver performance. Layout parasitic in configurable I/O driver are more critical when compared to other digital blocks like delay lines, driver controller and FSMs in PVT-compensation block, hence, this work only includes the configurable I/O driver in the layout implementation. **Figure 22** demonstrates the layout snapshot of the configurable I/O driver from **Figure 6**. The configurable I/O driver consists of three driver modules with each module divided into four driver stages. The total layout area is 312 × 196 μm2.

Also, during layout implementations, practical layout techniques like transistor splitting, symmetrical placement and transistor drain-source sharing are implemented into the I/O driver layout. This is because improper layout techniques in I/O driver can cause large layout parasitic that reduces the maximum operating speed of I/O

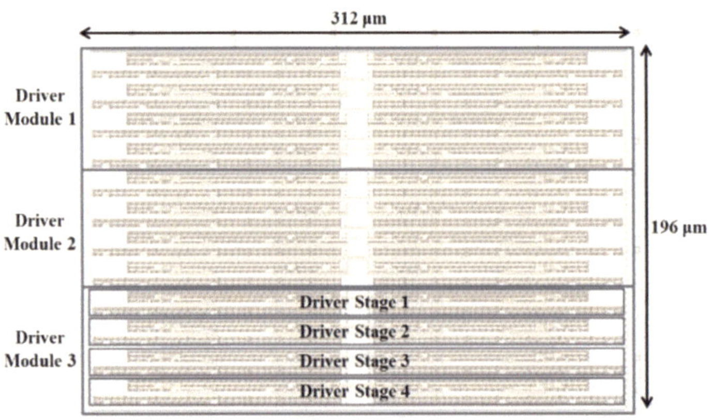

Figure 22.
*Layout of configurable I/O driver with three driver modules (**Figure 6**).*

driver. By applying transistor splitting technique, a single large transistor is split into smaller parallel transistors and this effectively reduces the parasitic resistance of transistor gate; thus, improving transistor switching speed for I/O driver. Additionally, differential current-mode driver is high performance analog circuit with matched devices in PMOS differential pair in driver stage (**Figure 10a**). Thus, it is necessary to place PMOS differential pair symmetrically for symmetric routing and matched parasitic [44]. Moreover, transistor drain-source sharing allows transistors to be placed as rectangle as possible that helps to compact and optimize layout area.

3.6 Test-bench setups

In this work, simulation flows are divided to two parts which are pre- and post-layout simulations. The circuit blocks in **Figure 6** are used as top-level test-bench for pre-layout performance verifications. After the configurable I/O driver is designed and simulated successfully, the I/O driver is implemented into layout. Similar to pre-layout simulations, the post-layout simulations also use the circuit in **Figure 6** as top-level test-bench, but the parasitic of I/O driver layout (see **Figure 22**) are extracted and loaded into PRESTO simulator to validate the design of configurable I/O driver especially for high data rate at 8 Gbps.

4. Post-layout results, discussions and challenges

This section discusses the post-layout simulation results in details. Simulation results of the output impedance and the output swing (current-mode driver) will be presented, followed by the discussions of the output slew rate and de-emphasis. Consequently, comparison and discussion between pre- and post-layout simulations will be presented.

4.1 On-chip calibration results

This section presents the on-chip calibration results for R-COMP and I-COMP across PVT variations. These PVT variations include the variation of process between fast, typical and slow corner; variation of supply voltage between 1.0 V ± 5%; and variation of temperature between 0, 50and 110°C.

4.1.1 R-COMP results

It is necessary to ensure that the driver stage impedance matched with a 600 Ω reference resistor to provide output impedance of 150 Ω per driver module. Thus, the proposed chip I/O design can be configured as output driver with output impedance of 50 Ω by enabling three driver modules during transmitting mode or configured as receive-ODT by enabling one, two or three driver modules to provide output impedance of 150, 75 or 50 Ω during receiving mode, respectively. **Figure 23a** and **b** shows the impedance range of pull-down and pull-up for one driver stage across PVT variations. Obviously, by controlling the four-bit binary-weighted dynamic legs, it is sufficient to control the driver stage pull-up and pull-down impedance at 600 Ω across PVT variations through the use of appropriate calibration code generated by the R-COMP.

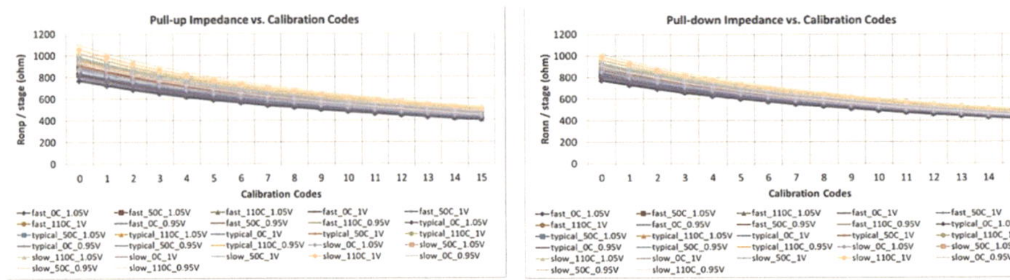

(a) (b)

Figure 23.
Driver stage impedance across calibration codes at different PVT corners for (a) pull-up impedance, and (b) pull-down impedance.

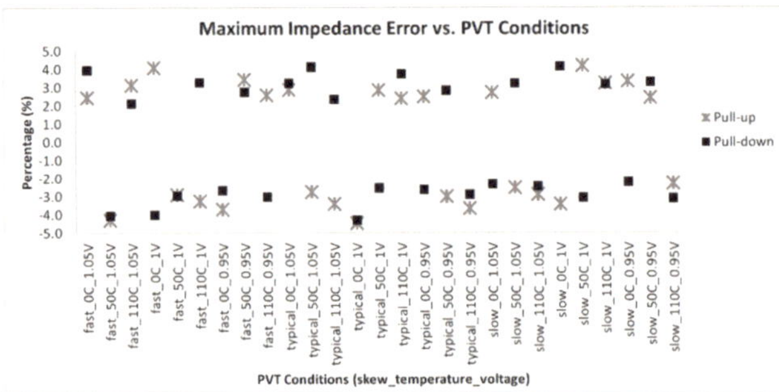

Figure 24.
Maximum impedance error of driver stage that is calibrated at 600 Ω across PVT variations.

It can be noticed that the impedance value in **Figure 23** is quantized across the calibration codes. R-COMP can select one of the two possible quantized impedances that are slightly higher or lower than 600 Ω. The difference between the quantized impedance and desired impedance is called impedance error. The percentage of maximum impedance error can be calculated. For instance, at typical condition (process: typical, voltage: 1.0 V and temperature: 50°C), calibration code 7 and code 8, provides pull-up impedance of 620.5 and 590.5 Ω, with percentage impedance error of 3.42 and −1.58% respectively. Moreover, **Figure 24** depicts the errors of maximum impedance for pull-up and pull-down impedance of the driver stage. It can be seen that the maximum impedance errors are less than ±5% across PVT variations and this is in agreement with the design requirements stated in **Table 1**. Besides, driver linearity is one of the design criteria in high speed communications. Ideally, the driver linearity should be 100% which implies that the output impedance is always constant across operating output voltages. Nevertheless, in reality the output impedance diverge from desired value across output voltages owing to non-linear characteristic of the transistor in both the pull-up and pull-down segment.

The graph of linearity errors of 600 Ω pull-up and pull-down impedance across operating voltages is shown in **Figure 25**. It can be noticed that the linearity errors are lower than the specification which is ±15% (see **Table 1**). Moreover, it can be seen that the pull-up impedance linearity error in **Figure 25a** is worst at low output voltages. This is because, the PMOS transistor in pull-up segment (see main driver in Section 3.4.1) is operating in non-linear region. This is the same reason for pull-down impedance worst linearity error at higher output voltages (**Figure 25b**) since the

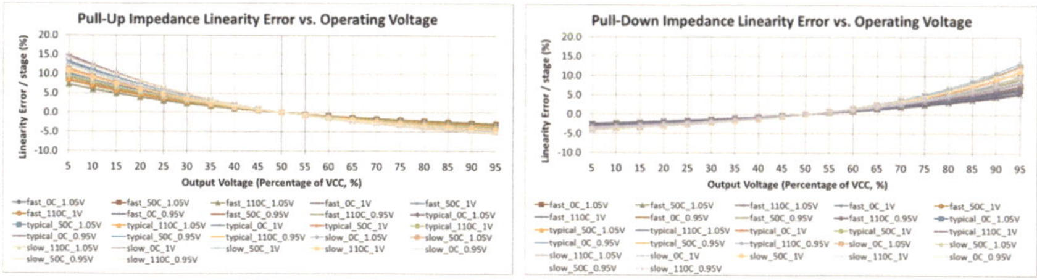

(a) (b)

Figure 25.
Linearity errors of 600 Ω across output voltages at different PVT corners for (a) pull-up impedance, and (b) pull-down impedance.

NMOS transistor in pull-down segment is not operating in linear region. In fact, linearity error can be reduced by increasing the series resistance against the transistor impedance in pull-up and pull-down segment in main driver in **Figure 7**. However, this requires larger transistor size that gives smaller transistor impedance to balance the increase of series resistance in order to keep output impedance at constant.

Moreover, larger transistor size can contribute to higher output pad capacitance that limits the driver data rates. Therefore, this is a compromise between linearity error and pad capacitance. Low linearity error across operating voltages helps to fully absorb any reflection from far-end but it also contributes to high pad capacitance that will slow down the slope of output and limit data transfer rates [45].

4.1.2 I-COMP results

When the proposed driver is configured as a current-mode driver, the output swing can be calibrated at desired level across different PVT corners. The impedance of the pull-down terminator of the current-mode driver is calibrated at 50 Ω (with three driver modules) through the calibration codes generated from R-COMP. **Figure 26a** and **b** shows the range of single-ended output swing of the current-mode driver over the calibration codes. Undoubtedly, by controlling the four-bit binary-weighted dynamic legs in the pull-up segment, the tail can be adjusted and calibrated at 400 and 500 mV with suitable calibration codes provided by I-COMP. It is critical to calibrate output swing at a fixed level throughout various corners to ensure the signal eye height is always sufficient for far-end receiver detection to avoid data sampling errors [24].

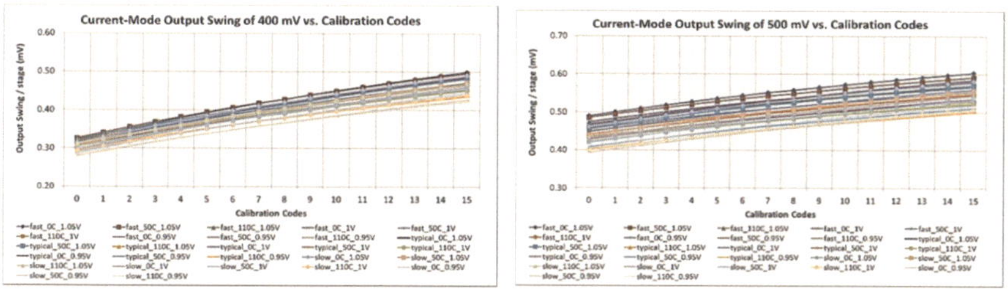

(a) (b)

Figure 26.
Output swing of current-mode driver can be calibrated by calibration codes for (a) 400, and (b) 500 mV.

Similar to output impedance discussed in the previous sub-section, it can be noticed that the output swing of current-mode driver is quantized across calibration codes (**Figure 26**) due to the discrete change of dynamic legs number. There are two possible quantized output swings which are slightly higher and lower than desired output swing. Thus, the difference between the quantized output swing and desired output swing can be called output swing error. The maximum output swing errors of less than ±5% across PVT variations were obtained and this fulfills the design requirement stated in **Table 1**.

4.1.3 Calibration results

On-chip impedance and output swing (in current-mode driver) are calibrated by adjusting the 4 bit-dynamic legs in the pull-up and pull-down segment of a driver stage. The impedance of pull-up segment is first calibrated, followed by impedance calibration of pull-down segment and output swing calibration (for current-mode driver). **Figure 27** shows the calibration sequences in R-COMP and I-COMP block for worst condition (slow transistor skew, 0.95 V supply voltage and 110°C temperature). The reference voltage, VREF, is half of the supply voltage i.e. 0.475 V. The waveform labels correspond to the voltage nodes in **Figures 16** and **17**. The driver pull-up impedance will be first calibrated by R-COMP (see **Figure 16a**). At t = 0 in **Figure 27a**, the VOUTP is lower than VREF which implies pull-up impedance is larger than reference resistor. This delta voltage will cause the comparator to output a high signal for U/D counter to increase the number of dynamic legs in the pull-up segment to rise the VOUTP voltage. When VOUTP is approaching and alternating about VREF, the PUN FSM will assert a lock signal LOCKP to hold the pull-up codes in the register at t = 70 ns, as shown in **Figure 27d**, to indicate that the pull-up impedance calibration process is completed by PUN FSM. In addition, this LOCKP signal also will kick-off the pull-down impedance calibration block with the pull-up segment and this becomes the reference impedance (see **Figure 16b**). At t = 70 ns in **Figure 27b**,

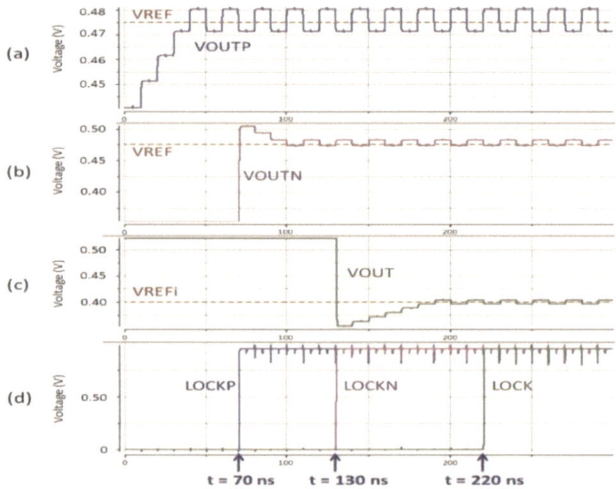

Figure 27.
Calibration process: (a) VOUTP in pull-up impedance calibration, (b) VOUTN in pull-down impedance calibration, (c) VOUT in output swing calibration, and (d) locks signal for pull-up, pull-down impedance and output swing.

VOUTN is higher than VREF and the dynamic legs in pull-down segment will be activated through the codes from U/D counter to lower the VOUTN. This search process will continue until VOUTN is alternating about VREF and the PDN FSM issues a lock signal LOCKN at t = 130 ns in **Figure 27d** as indication of calibration completed and to hold the pull-down codes in register. Once the pull-down impedance is calibrated with LOCKN signal asserted at t = 130 ns as shown in **Figure 27d**, the LOCKN will kick-off output swing calibration in I-COMP block (see **Figure 17**). Therefore, in **Figure 27c**, it can be noticed that the driver output VOUT becomes lower than output swing reference VREFi (e.g. 400 mV) after I-COMP block activated at t = 130 ns. This implies the current from pull-up segment is insufficient and the delta voltage will cause FSM in I-COMP to increase more dynamic legs to provide more current to raise up the VOUT approaching VREFi. When VOUT is alternating about VREFi, a lock signal LOCK is generated by FSM to store the codes into register at t = 220 ns, as shown in **Figure 27d**. This indicates that the calibration process is completed and all the codes that are held in register can be distributed to all the output drivers.

4.2 Output slew rate results

As discussed in Section 3.3 and 3.4.4.1, the delay cells in a delay line 1 and delay line 2 shall be designed to provide interval Td that is controlled at 40 ps across PVT variations to guarantee a controllable output slew rate at 2–4 V/ns. This can be achieved by controlling the 4-bit binary delay code DCODE[3:0] in the delay cell to control the delay across each delay cell at 40 ps. **Figure 28a** shows that the delay range across a delay cell can be controlled at 40 ps through the delay calibration code DCODE[3:0]. Similar to impedance and output swing, the change of delay is discrete across the calibration code and this causes the quantization error to alternate between 40 ps. The difference between quantized delay and desired delay is the delay error.

The percentages of maximum delay error are less than ±5% which shows the delays across each delay cell are well-controlled at 40 ps so that the output slew rates are controllable within 2–4 V/ns, as discussed in the following sub-sections.

4.2.1 Output slew rate of voltage-mode driver

By setting the appropriate delay calibration code to delay cells, the slew rates of configurable I/O driver are controlled at 2–4 V/ns for voltage-mode and current-mode driver configuration. **Figure 29a** shows the near-end output waveform of a differential voltage-mode driver (**Figure 9a**) with 50 Ω driving impedance and 50 Ω far-end terminator to ground, simulated at data rate of 1 Gbps in typical condition. In addition, it can be noticed that the output waveform is a sort of staircase, similar to

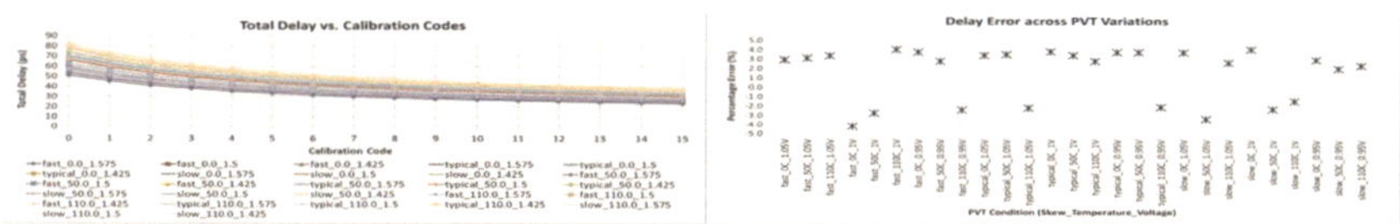

(a) (b)

Figure 28.
(a) Delay range of a delay cell with 4-bit dynamic transistor, and (b) percentage of maximum delay error from 40 ps across PVT variations.

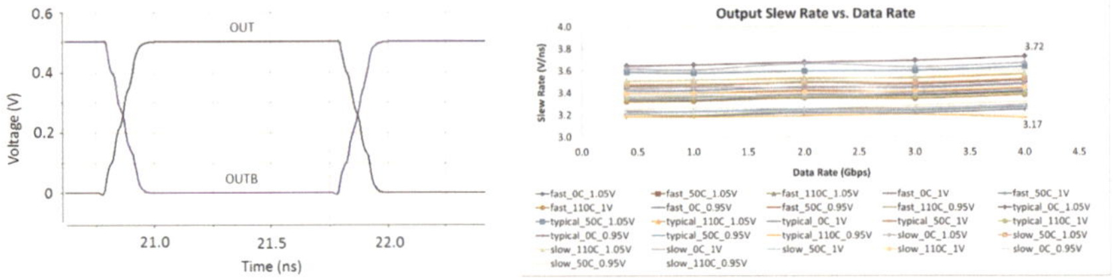

(a) (b)

Figure 29.
(a) Near-end output of differential voltage-mode driver, (b) near-end output slew rate of differential voltage-mode driver across PVT corners from 0.4–4.0 Gbps.

Figure 18b because each driver stage is activated in sequence to control the output transition time. Moreover, the output swing in **Figure 29a** is 500 mV. This resulted from voltage divider between near-end driver impedance and far-end terminator. Besides, **Figure 29b** shows the near-end output slew rates (measured from 20 to 80% of output swing) from 0.4–4.0 Gbps for different PVT corners. Obviously, output slew rates in **Figure 29b** can be controlled between 3.17–3.72 V/ns which meet the design requirements stated in **Table 1**. However, slew rate control scheme is bypassed to avoid restricting the output transition times for data rates above 4.0 Gbps (**Figure 19a**). Moreover, the output slew rate of single-ended driver (**Figure 8**) configured from the proposed configurable I/O driver is not shown in this sub-section due to the fact that both single-ended and differential voltage-mode driver are having same output slew rate, except that the output signaling is different in the form of single-ended or differential only.

Figure 30 shows the comparison between conventional voltage-mode driver without slew rate control and proposed driver with slew rate control simulated with package model. By enabling slew rate control, the proposed driver can reduce the maximum overshoot to 5.6% while the phase margin is improved from 50.1° to 63.8° and thus, giving a more stable and functional system. The reason to measure maximum overshoot at far-end is to ensure that the proposed I/O driver is capable of reducing ringing effect and avoid amplitude distortions that will cause sampling errors in the far-end receivers. Moreover, it can be observed that the slope of far-end output of the proposed driver become smooth in **Figure 30** as compared to its step-like near-end output in **Figure 29a**. This is because a lossy transmission channel has low-pass

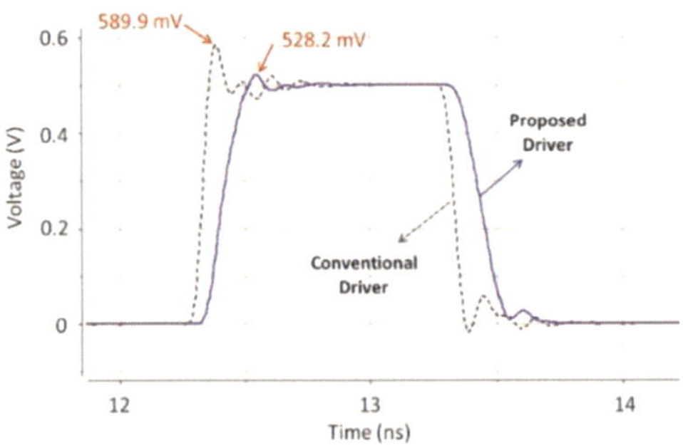

Figure 30.
Far-end output comparison between conventional voltage-mode driver and the proposed differential voltage-mode driver with slew rate control.

effect that filters the high-frequency components in the near-end output, thereby giving smoother slope of far-end output.

4.2.2 Output slew rate of current-mode driver

Differential current-mode driver in **Figure 10** also designed with controllable slew rate at 2–4 V/ns. **Figure 31a** and **b** show the near-end output waveforms of a differential current-mode driver with output swing of 400 and 500 mV (controlled by I-COMP block) respectively, simulated at 1 Gbps and typical condition. In addition, **Figure 32a** and **b** depict the near-end output slew rates (measured from 20 to 80% of output swing) for 400 and 500 mV output swing respectively, simulated from 0.4–4.0 Gbps for different PVT corners. The output slew rates in **Figure 32a** and **b** are controlled at 2.32–2.62 V/ns and 2.85–3.33 V/ns with calculated variation of 12.4 and 16.8%, respectively. Clearly, the output slew rates in **Figure 32a** and **b** can meet the targeted slew rate (**Table 1**). The slew rates in **Figure 32a** are generally smaller than that in **Figure 32b** due to smaller output swing since the output slew rate is the rate of change of output swing.

Moreover, **Figure 33** shows the comparison between conventional current-mode driver without slew rate control and proposed driver in current-mode with slew rate control simulated with package model. In **Figure 33a**, the 400 mV conventional current-mode driver has ringing with peak value at 476.9 mV on far-end output. Furthermore, the maximum overshoot and phase margin are calculated as 19.2% and 48.9°. **Figure 33a**, the proposed current-mode driver can reduce the peak ringing to 412.6 mV, while the maximum overshot and phase margin are improved to 3.15% and

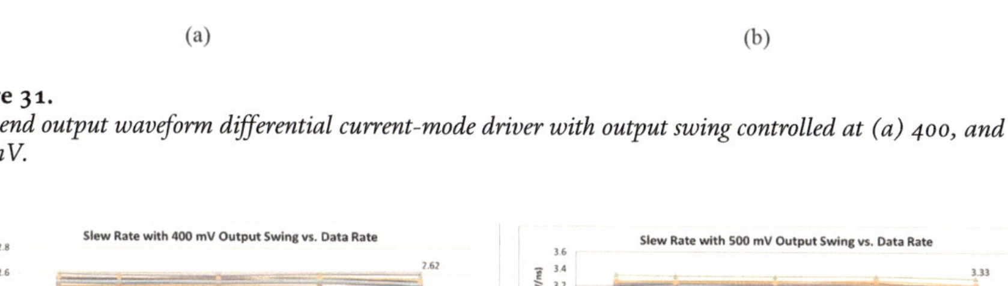

(a) (b)

Figure 31.
Near-end output waveform differential current-mode driver with output swing controlled at (a) 400, and (b) 500 mV.

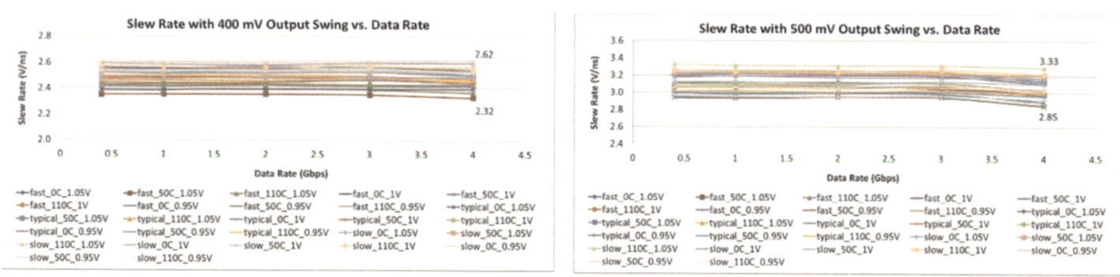

(a) (b)

Figure 32.
Near-end output slew rate of differential current-mode driver across PVT corners from 0.4–4.0 Gbps, with output swing of (a) 400, and (b) 500 mV.

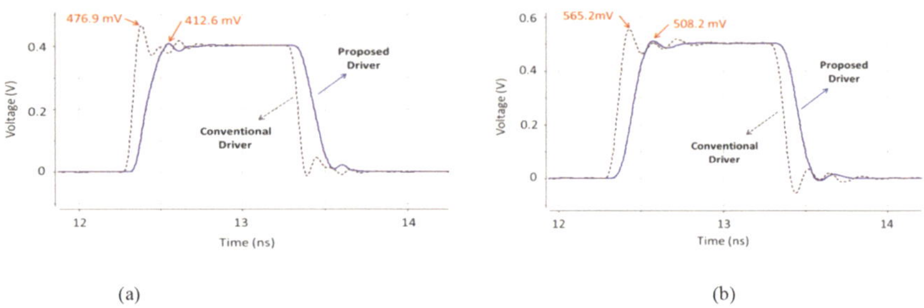

Figure 33.
Far-end output comparison between conventional driver and current-mode driver with slew rate control with output swing of (a) 400, (b) 500 mV.

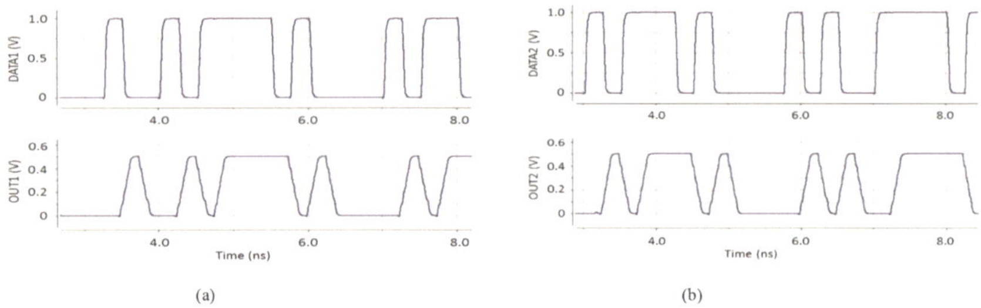

Figure 34.
Near-end output waveform of two single-ended voltage-mode drivers at OUT1 pin, (b) OUT2 pin.

67.2°, respectively. Also, in **Figure 34b**, there is peak ringing of 565.2 mV on the far-end output from a 500 mV conventional current-mode driver with the corresponding maximum overshoot and phase margin computed as 13.0% and 55.2°. The current-mode driver configured from the proposed driver with output swing of 500 mV is capable of reducing the far-end peak ringing and maximum overshoot to 508.2 mV and 1.64% and hence, giving better phase margin at 69.6° as compared to that from conventional current-mode driver. Also, from the results shown in **Figure 33a** and **b**, it can be noticed that the smaller output swing and slower slew rate are better in reducing output ringing. However, for data rate above 4.0 Gbps, slew rate control should be bypassed as slower slew rate is limiting driver data rates.

4.3 Output de-emphasis results

4.3.1 Voltage-mode de-emphasis

The proposed configurable I/O driver can support wide range of data rates from 0.4 to 8 Gbps, as stated by the specifications in **Table 1**. Single-ended voltage-mode driver (**Figure 8**) is designed to support data rates from 0.4 up to 4.0 Gbps, while differential voltage-mode driver (**Figure 9**) can support data rates from up to 8.0 Gbps due to the merit of differential signaling. **Figure 34** shows two near-end output waveforms when the proposed configurable I/O driver is configured as two single-ended drivers with 50 Ω driving impedance and 50 Ω far-end terminator, simulated at 4.0 Gbps in typical condition with slew rate control scheme activated.

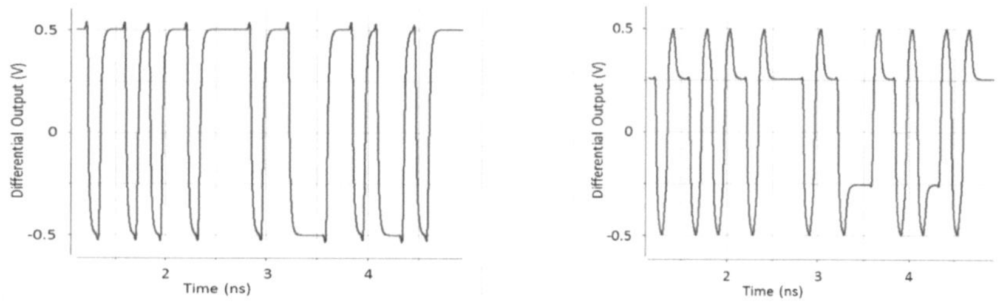

(a) (b)

Figure 35.
Differential output waveform of voltage-mode driver at 8 Gbps (a) without de-emphasis, and (b) with de-emphasis.

Moreover, 6 dB de-emphasis control scheme is programmable and can be activated for long transmission channel and higher data rates (e.g. above 4.0 Gbps) since the frequency dependent attenuations of transmission channel become significant and cause ISI problem. **Figure 35a** and **b** demonstrate the differential output waveform of voltage-mode driver without and with de-emphasis. The differential voltage-mode driver is simulated at 8 Gbps in typical condition with 50 Ω driving impedance and 50 Ω far-end pull-down terminator.

4.3.2 Current-mode de-emphasis

Differential current-mode driver (**Figure 10**) configured from the proposed configurable I/O driver also supports de-emphasis scheme up to 8.0 Gbps. **Figure 36a** shows the differential output waveform of current-mode driver with single-ended swing calibrated at 400 mV while **Figure 36b** shows the differential output waveform of current-mode driver with 6 dB de-emphasis at 8 Gbps in typical condition.

Figure 37a shows the differential output waveform of 500 mV current-mode driver while **Figure 37b** shows the differential output of 500 mV current-mode driver with 6 dB de-emphasis at 8 Gbps in typical condition.

4.4 I/O driver de-emphasis performance at high data rates

The ISI is frequency and transmission channel length dependent. Thus, post-layout simulations in this work are focusing on maximum data rate at 8 Gbps. Moreover, two

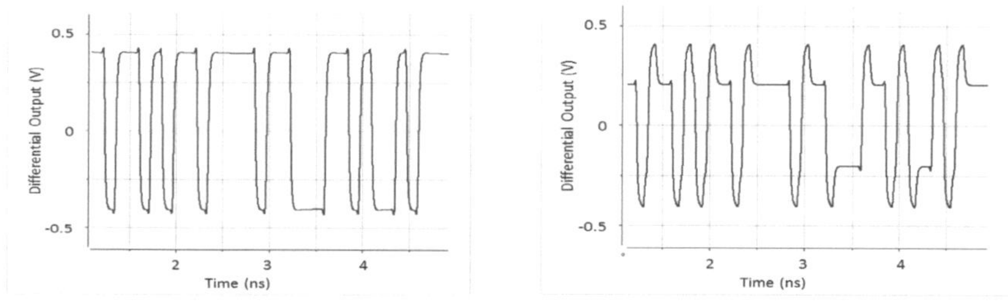

(a) (b)

Figure 36.
Differential output waveform of 400 mV current-mode driver at 8 Gbps (a) without de-emphasis, and (b) with de-emphasis.

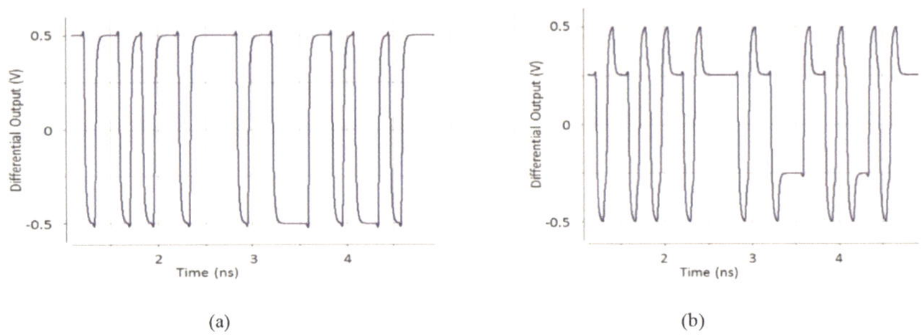

Figure 37.
Differential output waveform of 500 mV current-mode driver at 8 Gbps (a) without de-emphasis, and (b) with de-emphasis.

typical short and long transmission channel length in MCM (10 mm and 50 mm) will be used to verify the de-emphasis performance. At short channel, ISI impacts are not significant and the de-emphasis scheme (Section 3.4.4.2) can be bypassed. However, as ISI impacts become dominant at longer channel, de-emphasis scheme can be activated to equalize the channel attenuations, thereby improving far-end signal integrity as discussed in the following sub-sections.

4.4.1 Output driver performance at short transmission channels

When the proposed configurable I/O driver is placed in the package with short transmission channels i.e. 2.5–10 mm, de-emphasis scheme is bypassed since short channel will not lead to high attenuation on outputs. **Figure 38** shows the far-end differential eye signal of voltage-mode and current-mode driver in typical condition, simulated at 8 Gbps with 10 mm transmission channel. In **Figure 38a**, voltage-mode driver has eye jitter and eye height of 0.08 UI and 793.8 mV. Moreover, **Figure 38b** shows the 500 mV current-mode driver has eye jitter and eye height of 0.06 UI and 850.6 mV, while **Figure 38c** shows the eye jitter and eye height of 400 mV current-mode driver measured at 0.06 UI and 687.2 mV.

It can be seen that all the three drivers have sufficient inner eye opening that are larger than the specifications stated in **Table 1**, where 500 mV current-mode driver has better eye jitter and height performance.

4.4.2 I/O driver performance at long interconnects

When the proposed configurable I/O driver is operated with longer transmission channel up to 50.0 mm, the de-emphasis control scheme has to be activated to allow

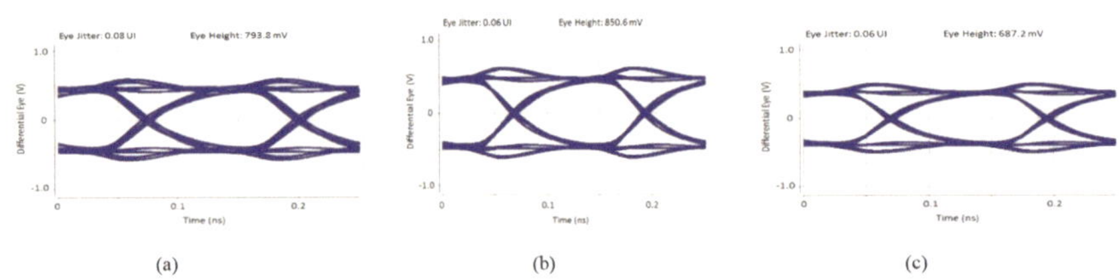

Figure 38.
Far-end differential eye without de-emphasis, simulated for (a) voltage- mode driver, (b) 500 mV current-mode driver, and (c) 400 mV current-mode driver.

adequate eye opening at high data rates. **Figure 39a** shows the far-end differential eye diagram of voltage-mode driver without applying de-emphasis at 8 Gbps in typical condition. The eye jitter and eye height are 0.50 UI and 226.9 mV respectively. Large eye jitter due to long transmission channel can cause receiver sampling errors.

Therefore, de-emphasis scheme should be activated to equalize the attenuations of transmission channel. **Figure 39b** depicts the far-end differential eye diagram of voltage-mode driver simulated at 8 Gbps with de-emphasis. The eye jitter and eye height are improved to 0.29 UI and 312.8 mV, which are large enough for timing and amplitude budgets of far-end receiver (**Table 1**).

Figure 40a shows the far-end differential eye diagram of 500 mV current-mode driver at 8 Gbps, typical condition and 50 mm transmission channel without applying de-emphasis. The eye jitter and eye height are 0.47 UI and 316.7 mV which are better than that in voltage-mode driver without de-emphasis (**Figure 39a**). By applying 6 dB of de-emphasis, the far-end differential eye opening is improved with eye jitter and eye height of 0.26 UI and 367.5 mV respectively, as demonstrated in **Figure 40b**. Thus, after applying de-emphasis, the eye jitter and eye height are improved to meet the timing and amplitude budgets of far-end receiver, as mentioned in **Table 1**.

Figure 41a shows the far-end differential eye diagram of 400 mV current-mode driver simulated with 8 Gbps and 50 mm transmission channel without de-emphasis. The eye jitter and eye height are 0.46 UI and 254.3 mV. By applying de-emphasis, the far-end differential eye opening is improved with eye jitter and eye height of 0.27 UI and 277.8 mV respectively, as shown in **Figure 42b**.

In **Figures 39** and **40**, it can be noticed that 500 mV current-mode driver has better eye jitter and eye height performance as compared to voltage-mode driver due to the fact that current-mode driver has faster transient response, as shown in

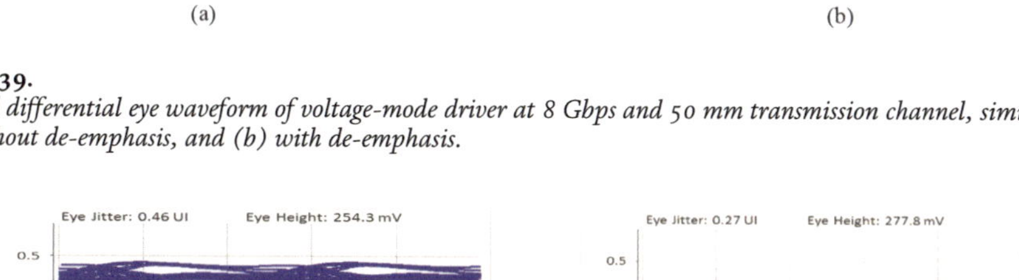

(a) (b)

Figure 39.
Far-end differential eye waveform of voltage-mode driver at 8 Gbps and 50 mm transmission channel, simulated (a) without de-emphasis, and (b) with de-emphasis.

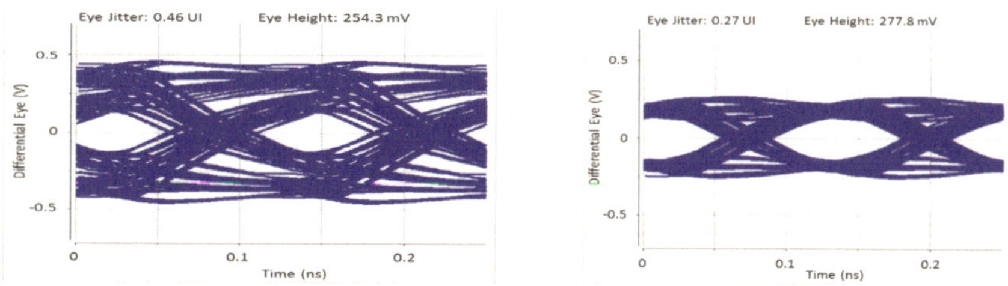

(a) (b)

Figure 40.
Far-end differential eye of 500 mV current-mode driver at 8 Gbps and 50 mm transmission channel, simulated (a) without de-emphasis, and (b) with de-emphasis.

(a) (b)

Figure 41.
Far-end differential eye of 400 mV current-mode driver at 8 Gbps and 50 mm transmission channel, simulated (a) without de-emphasis, and (b) with de-emphasis.

Figure 42a. Generally, the output voltage of both the drivers at high data rates depend on the charging and discharging of the capacitance at internal and output nodes. Current-mode driver has stable source voltages VS1 and VS2 where the node capacitance is always charged-up by tail currents to constant voltage level, as shown in **Figure 42b**. For instance, when output node OUT is driving a logic high (when data D is low), a total current of $2I_b$ is charging the capacitance at the node between transistor M3 and resistor R1 and output node. This reduces transistor turned-on time as compared to voltage-mode driver where every single node capacitance including output node has to be charged-up along the pull-up branch. Thus, current-mode driver has faster transient response, giving larger eye opening when compared to voltage-mode driver. To avoid sampling errors for the applications that require low jitter, it is suggested to configure the proposed output driver into current-mode driver at the expense of slightly higher power consumption as shown in the graph of average power performance in **Figure 43**.

In **Figures 40** and **41**, the 500 mV current-mode driver has similar eye jitter but higher eye height as compared to 400 mV current-mode driver. The eye jitter is almost identical because both of them has same driving impedance and far-end terminators that matches the transmission channel characteristic impedance. Thus, no additional jitters due to signal reflections for both of the current-mode drivers. However, 500 mV current-mode driver has higher eye height due to its higher output swing. Therefore, for high performance applications that need higher voltage margin or eye height to avoid wrong data logic detection, it is suggested to configure the proposed I/O driver into 500 mV current-mode driver instead as 400 mV

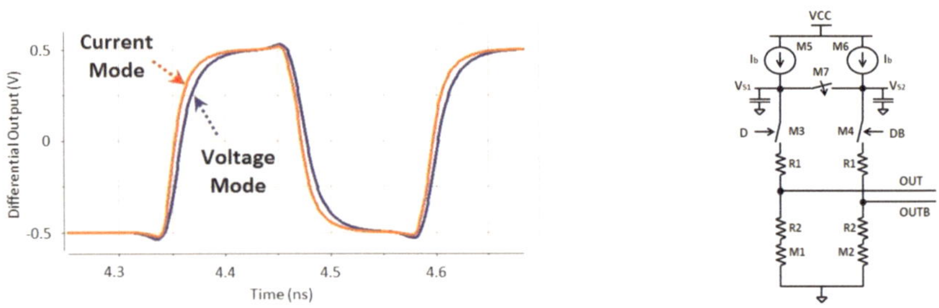

(a) (b)

Figure 42.
(a) Comparison of output transient response between voltage-mode driver and 500 mV current-mode driver, simulated at 8 Gbps in typical condition, and (b) current-mode driver with stable source voltage at VS1 and VS2.

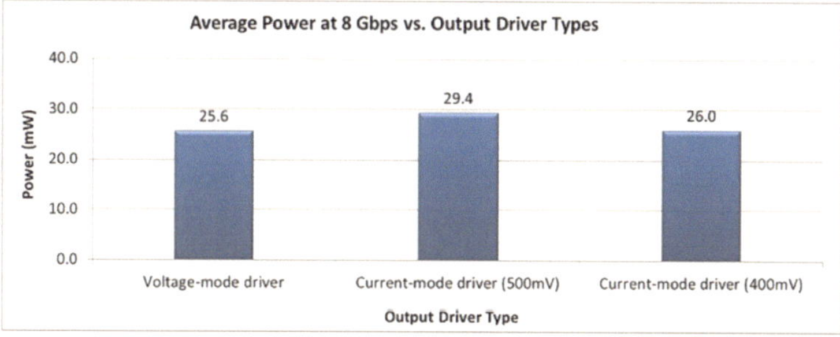

Figure 43.
Comparison of average power between voltage-mode driver, 400 and 500 mV current-mode driver, simulated at 8 Gbps and typical condition.

current-driver, at the expense of slightly higher power consumption. If voltage margin is not a concern, then the 400 mV current-mode driver should be configured to avoid overhead of power consumption (**Figure 43**).

4.4.3 ODT performance with 50, 75 and 150 Ω impedance

When receiving data from the far-end driver, the proposed configurable I/O driver can be configured as near-end receive ODT with 50, 75 or 150 Ω by enabling the corresponding number of driver modules, as discussed in Section 3.4.1.3. **Figure 44a** shows the near-end eye diagram of 50 Ω ODT when it terminates the output transmitted by far-end 8 Gbps I/O driver through 10 mm of transmission channel. The 50 Ω ODT is well-matched to the transmission channel impedance and the simulated jitter and eye height are 0.10 UI and 571.2 mV, respectively.

Consequently, 75 and 150 Ω ODT can be configured as near-end termination for power saving purposes at the expense of eye jitter, because 75 and 150 Ω ODT are not matched with transmission channel impedance and this causes reflection noise, as shown by the eye diagram in **Figure 44b** and **c**. Eye jitter is increased to 0.18 UI and 0.29 UI for 75 and 150 Ω ODT, respectively. However, eye heights are improved to 612.1 and 633.2 mV for 75 and 150 Ω ODT due to the fact that the output swing increases with larger ODT impedance. Nevertheless, in overall, the eye opening is still sufficient for data detection and sampling. In addition, **Figure 45** demonstrates the graph of average power consumption for ODT impedance of 50, 75 and 150 Ω, simulated at 8 Gbps. It can be noticed that the average power of near-end ODT is reducing with larger ODT impedance. Hence, 75 and 150 Ω ODT can be configured for low power applications, while 50 Ω ODT should be selected to terminate reflection noises in high performance applications.

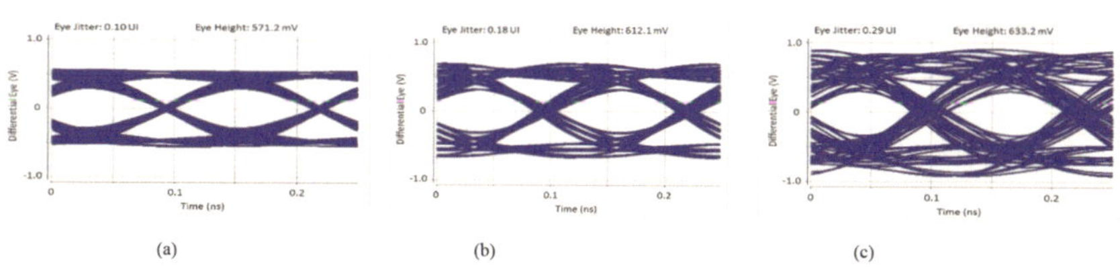

Figure 44.
Near-end signal eye with ODT value of (a) 50, (b) 75, and (c) 150 Ω, simulated at 8 Gbps and transmission channel length of 10 mm.

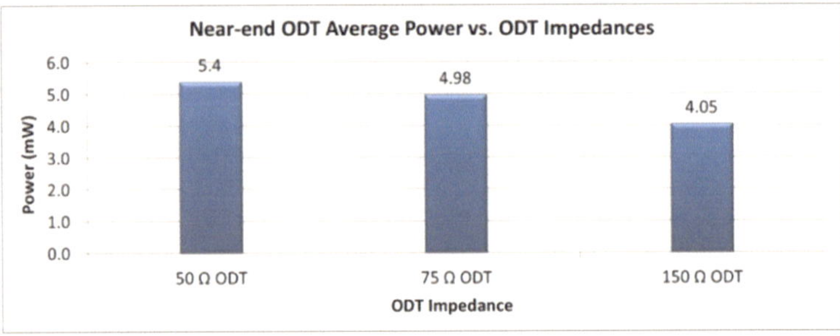

Figure 45.
Average power consumption of ODT value of 50, 75, and 150 Ω.

4.5 Performance summary

This section summarizes and compares the simulated signal eye jitter and eye height between pre-layout and post-layout simulations with 50 mm transmission channel. **Figure 46a** demonstrates the eye jitters measured from pre-layout and post-layout for differential voltage-mode and current-mode driver with de-emphasis, simulated with data rate of 8 Gbps and 50 mm transmission channel in typical condition. Also, **Figure 46b** depicts the pre-layout and post-layout simulated eye height (8 Gbps and typical condition) for voltage-mode and current-mode drivers. Thus, the eye jitter and eye height show good agreement between pre- and post-layout since the variations between them are less than 4%.

Table 4 shows the performance summary of the proposed configurable output driver. Different impedance value i.e. 50, 75 or 150 Ω can be selected for near-end ODT. Overall, the output slew rates of differential voltage-mode driver and current-mode driver can be controlled within targeted value at 2–4 V/ns. Moreover, after applying de-emphasis on the differential outputs at 8 Gbps, it can be found that the current-mode driver has better eye jitter and eye height performance when compared to voltage-mode driver. Although current-mode driver has better eye opening at 8 Gbps, but its average power consumption is slightly higher than that in the voltage-mode driver. This shows the performance trade-off between signal eye opening and power consumption. Current-mode driver can be configured for high performance signaling system due to its advantage of reducing ISI impacts at the cost of slightly higher power consumption. On the other hand, voltage-mode driver is suggested for low power system due to the merit of lower power consumption as compared to current- mode driver.

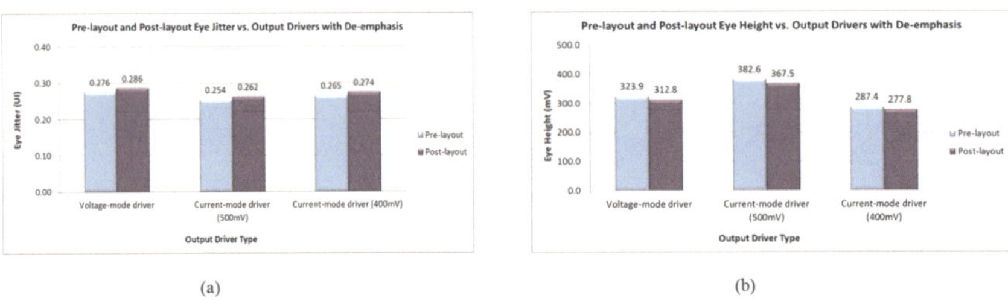

Figure 46.
Pre-layout and post-layout simulations of voltage-mode and current-mode drivers on (a) eye jitters, and (b) eye heights.

		Specifications in Table 3	Voltage-Mode	Current-Mode (500 mV)	Current-Mode (400 mV)
Process technology		—	45 nm CMOS		
Silicon area (three driver modules)			$312 \times 196\ \mu m^2$		
Operating speed			0.4–8 Gbps		
Operating VCC			1.0 V ± 5%		
Worst impedance error (%)		± 5 Max.	4.76		
Worst impedance linerity error (%)		± 15 Max.	14.80		
Worst output swing error (%)		± 5 Max.	—	−0.96	1.49
Output slew rate (V/ns)		2.0–4.0	3.17–3.72	2.85–3.33	2.33–2.62
Typical average power (mW)	50 Ω ODT	—	5.40		
	75 Ω ODT		4.98		
	150 Ω ODT		4.05		
	50 Ω Driver		25.64	29.36	26.03
Driver without de-emphasis (at 8 Gbps, 50 mm of TL)	Eye jitter (UI)	0.30 Max.	0.50	0.47	0.46
	Eye height (mV)	200.0 Min.	226.90	316.70	254.30
Driver with de-emphasis (at 8 Gbps, 50 mm of TL)	Eye jitter (UI)	0.30 Max.	0.29	0.26	0.27
	Eye height (mV)	200.00 Min.	312.80	367.50	277.80
Signal eye improvement with 6 dB de-emphasis	Eye jitter (%)	—	42.32	44.68	41.30
	Eye height (%)		37.86	16.04	9.24

Table 4.
Performance summary of configurable I/O driver at 8 Gbps.

4.6 Performance comparison with other works

The performance comparison between the proposed configurable I/O driver and other output driver results from literatures is shown in **Table 5**.

From **Table 5**, the 500 and 400 mV current-mode driver configured from the proposed configurable I/O driver have better output swing control with variation of −0.96 and 1.49% respectively, as compared to 8.2% swing variation in the output driver presented by Ref. [39]. Moreover, by controlling the output swing at almost constant level, the output slew rate is automatically controlled between 3.11–3.77 V/ns with variation of 21.2% as summarized in **Table 5** [39]. Yet, the slew rate control scheme in this work allows the slew rates of the configurable I/O driver to become less sensitive to PVT variations. The slew rate of voltage-mode driver, 500 mV current-mode driver and 400 mV current-mode driver has smaller PVT variation at 17.35, 16.84 and 12.45% respectively.

From **Table 5**, the transmitter in Ref. [46] has better capability of improving 67.8% of eye height when compared to the drivers from proposed configurable I/O driver even though the eye height of each driver is larger than 200 mV. In addition, it can be noticed that the eye jitters of voltage-mode driver, 500 mV current-mode driver and 400 mV current-mode driver without de-emphasis are larger than 0.41 UI

		This work			References	
		Voltage-mode driver	Current-mode driver		[39]	[46]
Process technology		45 nm CMOS			65 nm CMOS	90 nm CMOS
Supply voltage (V)		1.0			1.0	1.2
Driver mode		Multi-mode			Single mode	Single mode
Data rate (Gbps)		0.4–8.0			10.0	2.0–6.0
Output swing (mV) output		500	500	400	400	100–400
Swing control		None	Yes	Yes	Yes	None
Output swing variation (%)		None	−0.96	1.49	8.20	—
Slew rate (V / ns)		3.17–3.72	2.85–3.33	2.33–2.62	3.11–3.77	—
Slew rate variation (%)		17.35	16.84	12.45	21.2	—
Driver de-emphasis		Yes	Yes	Yes	None	Yes
Without	Eye jitter (UI)	0.50	0.47	0.46	—	0.41
De-emphasis	Eye height (mV)	226.9	316.7	254.3	—	87.0
With	Eye jitter (UI)	0.29	0.26	0.27	—	0.28
De-emphasis	Eye height (mV)	312.8	367.5	277.8	—	146.0
Eye opening	Eye jitter (%)	42.32	44.68	41.3	—	31.71
Improvement	Eye height (%)	37.86	16.04	9.24	—	67.82

Table 5.
Performance comparison with other published works.

in Ref. [46] due to the use of different transmission channel length. However, the configurable I/O driver can reduce 42.3, 44.7 and 41.3% of eye jitter for the corresponding voltage-mode driver, 500 mV current-mode driver and 400 mV current-mode driver, respectively, when compared to 31.7% in Ref. [46]. This shows that the proposed configurable I/O driver with de-emphasis control scheme has better capability in eye jitter reduction at high data rates.

5. Conclusion

In this work, we have demonstrated the fundamentals, methodologies, and challenges of chip I/O design with the design of a configurable I/O driver supporting wide range of data rates from 0.4 to 8.0 Gbps. The proposed configurable I/O driver can be configured between two single-ended voltage-mode drivers, a differential voltage-mode driver and a differential current-mode driver to address electronic signaling mismatches between chips, thereby achieving maximum performance across a variety of processor-intended applications. Moreover, four control schemes, namely, impedance control, output swing control, slew rate control and de-emphasis control scheme are embedded into configurable I/O driver design to improve signal integrity especially at high data rates. Through the impedance control scheme, output impedance is calibrated at 50 Ω for driver as well as 50, 75 and 150 Ω for ODT, with worst impedance error at 4.76%. The output swing of current-mode driver can be controlled by output swing control scheme at 400 and 500 mV with worst output swing errors of −0.96 and 1.49% respectively. Besides, the slew rate control scheme can slow down the slew rate of configurable I/O driver to 2–4 V/ns to address switching noises. The output slew rate of voltage-mode driver, 400 mV current-mode driver and 500 mV current- mode driver can be controlled at 3.17–3.72 V/ns, 2.33–2.62 V/ns and 2.85–3.33 V/ns, respectively. With de-emphasis control scheme, the voltage-mode driver, 500 mV current-mode driver and 400 mV current-mode driver are capable of reducing far-end eye jitter by about 42.3, 44.7 and 41.3%. The 500 mV current-mode driver has relatively larger far end eye opening, but it consumed extra 12.7% of average power as compared to voltage-driver. Thus, with the four digital schemes mentioned above that forms the auxiliary control, the proposed configurable I/O driver is capable of supporting multiple signaling types while improving signal integrity at high frequency data transmissions, thereby enhancing chip-to-chip performance to boost the performance of electronic devices like computers and mobile phones for better consumer experiences.

Author details

Ikechi A. Ukaegbu[1*] and Arjuna Marzuki[2]

1 Electrical and Computer Engineering Department, School of Engineering and Digital Sciences (SEDS), Nazarbayev University, Astana, Kazakhstan

2 School of Science and Technology (SST), Wawasan Open University, Penang, Malaysia

*Address all correspondence to: ikechi.ukaegbu@ieee.org

References

[1] Badawy W, Jullien GA. System-on-Chip for Real-Time Applications. Boston: Kluwer Academic Publishers; 2003

[2] Balch M. Complete Digital Design - A Comprehensive Guide to Digital Electronics and Computer System Architecture. New York, NY: McGraw-Hill; 2003

[3] Tsuchiya A, Kuboki T, Onodera H. Low-power design of CML drivers for on-Chip transmission-lines. IEICE Transactions on Electronics. 2007;**6**: 1274-1281

[4] Ukaegbu IA, Kim D-W, Cho MH, Lee TW, Park H-H. 2.5 Gb/s/ch long wavelength transmitter modules for Chip-to-chip Optical PCB applications. IEEE Photonics Technology Letters. 2011;**23**(19):1403-1405

[5] Ukaegbu IA, Kim D-W, Cho MH, Lee TW, Park H-H. Performance analysis of vertical and horizontal transmitter Array modules using short and long wavelength VCSELs for optical interconnects. IEEE Transactions on Components, Packaging and Manufacturing Technology. 2013;**3**(5): 740-748

[6] Warnock J, Chan YH, Harrer H, Rude D, Puri R, Carey S, et al. 5.5 GHz system Z microprocessor and multi-chip module. Proceeding of IEEE International Solid-State Circuits Conference Digest of Technical Papers (ISSCC). 2013:46-47

[7] Robertson ID, Lucyszyn S. RFIC and MMIC Design and Technology. London, U.K.: The institution of Electrical Engineers; 2009

[8] Dally WJ, Poulton J. Digital Systems Engineering. Cambridge, U.K.: Cambridge University Press; 2001

[9] Hussain W, Valorge O, Blaquière Y, Savaria Y. A novel spatially configurable differential interface for an electronic system prototyping platform. VLSI Journal of Integration. 2016;**55**:129-137

[10] Bogatin E. Roadmaps of Packaging Technology. Scottsdale, AR: Integrated Circuit Engineering Corporation; 1997

[11] Wang S, Tao Y, Bereza W, Kwasniewski T. US Patent No. 7,598,779. Washington, DC: U.S. Patent and Trademark Office; 2009

[12] Chengson DP, Conrad RA. U.S. Patent No. 5,811,977. Washington, DC: U.S. Patent and Trademark Office; 1998

[13] Song HJ. Dual mode transmitter with adaptively controlled slew rate and impedance supporting wide range data rates. Proceeding of 14th Annual IEEE International ASIC/SOC Conference. 2001:321-324

[14] Gui X, Tang R, Li K, Wang K, Li D, Pan Q, et al. A CMOS slew-rate controlled output driver with low process, voltage and temperature variations using a dual-path signal-superposition technique. IET Circuits, Devices and Systems. 2023;**17**(1):13-28

[15] Heidar D, Dessouky M, Ragaie HF. Comparison of output drivers for high-speed serial links. Proceeding of IEEE International Conference on Microelectronics (ICM). 2007:329-332

[16] Celit F. Energy-Efficient Design Techniques for High-Speed Wireline Serial Links. Switzerland: EPFL Scientific Publications; 2021

[17] Higashi H, Masaki S, Kibune M, Matsubara S, Chiba T, Doi Y, et al. A

5-6.4- Gb/s 12-channel transceiver with pre-emphasis and equalization. IEEE Journal of Solid-Stage Circuits. 2005;**40**: 978-985

[18] Chen W, Cai Q, Yang Z, Qiao S. A 0.8–6Gb/s wireline receiver based on the spectrum-balancing equalizer and semi-digital dual loop CDR. IEICE Electronics Express. 2023;**20**(10):1-6

[19] Bartolini M, Pulici P, Stoppino PP, Campardo G, Vanalli GP. A reduced output ringing CMOS buffer. IEEE Transactions on Circuits and Systems II: Express Briefs. 2007;**54**:102-106

[20] Micheloni R, Crippa L, Marelli A. Inside NAND Flash Memories. New York: Springer; 2010

[21] Lin Y, Zhang J, Zou XL, D Wang S-Y. Design of output buffer with low switching noise and load adaptability. Analog Integrated Circuit Signal Process. 2010;**65**:273-282

[22] Esch GJ, Chen T. Near-linear CMOS I/O driver with less sensitivity to process, voltage, and temperature variations. IEEE Transactions on Very Large Scale Integration (VLSI) Systems. 2004;**12**:1253-1257

[23] Fan Y, Smith JE. On-die termination resistors with Analog impedance control for standard CMOS technology. IEEE Journal of Solid-Stage Circuits. 2003;**38**: 361-364

[24] Muljono H, Tian K, Atha M, Lin C, Sun LK, Rusu S. A 6.4 GT/s point-to-point unidirectional link with full current compensation. Proceeding of IEEE Asian Solid-State Circuits Conference. 2009:169-172

[25] Norman R, Valorge O, Blaquiere Y, Lepercq E, Basile-Bellavance Y, El-Alaoui Y, et al. An Active Reconfigurable Circuit Board. Proceeding of International IEEE Conference on Circuits and Systems and TAISA. 2008: 351-354

[26] Kossel M, Menolfi C, Weiss J, Buchmann GvB, Rodoni L, Morf T, et al. A T-coil-enhanced 8.5 Gb/s high-swing SST transmitter in 65nm bulk CMOS with <-16dB return loss over 10GHz bandwidth. IEEE Journal of Solid-State Circuits. 2008;**43**(12):2905-2919

[27] Li H, He L, Xu X, Yan X. A dual mode serial link transmitter with controlled slew rate. Proceeding of IEEE International Conference on Solid-Stated and Integrated Circuits Technology. 2004;**2**:1575-1578

[28] Jou SJ, Kuo SH, Chiu JT, Lin TH. Low switching noise and load adaptive output buffer design techniques. IEEE Journal of Solid-Stage Circuits. 2001;**36**: 1239-1249

[29] Hall SH, Heck HL. Advanced Signal Integrity and Digital High Speed Design. Hoboken, NJ: John Wiley and Sons; 2009

[30] Gabara TJ, Knauer SC. Digitally adjustable resistors in CMOS for high-performance applications. IEEE Journal of Solid-State Circuits. 1992;**27**:1176-1185

[31] DeHon A, Knight T, Simon T. Automatic Impedance Control. Proceeding of IEEE International Solid-State Circuits Conference. 1993:164-165

[32] Zhang L, Wilson JM, Bashirullah R, Luo L, Xu J, Franzon PD. Voltage-mode driver Preemphasis technique for on-chip global buses. IEEE Transactions on Very Large Scale Integration (VLSI) Systems. 2007;**15**:231-236

[33] Liu J, Lin X. Equalization in high-speed communication systems. IEEE

Circuits and Systems Magazine. 2004;**4**: 4-17

[34] Li M, Kwasnieski T, Wang S, Tao Y. A 10Gb/s transmitter with multi- tap FIR pre-emphasis in 0.18μm CMOS technology. Proceeding of IEEE on Design Automation Conference. 2005;**2**: 679-682

[35] Kim H, Cho J, Kim M, Kim K, Lee J, Lee H, et al. Measurement and analysis of a high-speed TSV Channel. IEEE Transactions on Components, Packaging, and Manufacturing Technology. 2012;**2**:1672-1685

[36] Kim JH, Oh D, Kim W. Accurate characterization of broadband multiconductor transmission channels for high-speed digital systems. IEEE Transactions on Advanced Packaging. 2010;**33**:857-867

[37] Hall SH, Hall GW, McCall JA. High-Speed Digital System Design: A Handbook of Interconnect Theory and Design Practices. NY: John Wiley and Sons; 2000

[38] Kim JK, Kim J, Jeong DK. A 20-Gb/s full-rate 2^7-1 PRBS generator integrated with 20-GHz PLL in 0.13-μm CMOS. Proceeding of IEEE Asian Solid-State Circuits Conference (ASSCC). 2008: 221-224

[39] Kim T, Jeong DK. A 10 Gb/s voltage swing level controlled output driver in 65-nm CMOS technology. Proceeding of International IEEE in SoC Design Conference 2012 (ISOCC). 2012: 53-56

[40] Kalyan G, Srinivas MB. An efficient ODT calibration scheme for improved signal integrity in memory Interface. Proceeding of IEEE Asia Pacific Conference on Circuits and Systems (APCCAS). 2010:1211-1214

[41] Jung HK, Lee K, Kim JS, Lee JJ, Sim JY, Park HJ. A 4 Gb/s 3-bit parallel transmitter with the crosstalk-induced jitter compensation using TX data timing control. IEEE Journal of Solid-State Circuits. 2009;**44**:2891-2900

[42] Maymandi NM, Sachdev M. A monotonic digitally controlled delay element. IEEE Journal of Solid-State Circuits. 2005;**40**:2212-2219

[43] Menolfi C, Toifl T, Buchmann P, Kossel M, Morf T, Weiss J, et al. A 16 Gb/s source-series-terminated transmitter in 65 nm CMOS SOI. Proceeding of IEEE International Solid-State Circuits Conference (ISSCC). 2007:446-447

[44] Lampaert K, Gielen G, Sansen W. Analog Layout Generation Performance and Manufacturability. Boston: Kluwer Academic Publishers; 2010

[45] Park C, Chung H, Lee YS, Kim J, Lee JJ, Chae MS, et al. A 512-Mb DDR3 SDRAM prototype with CIO minimization and self-calibration techniques. IEEE Journal of Solid-State Circuits. 2006;**41**:831-838

[46] Song YH, Palermo S. A 6-Gbit/s hybrid voltage-mode transmitter with current-mode equalization in 90-nm CMOS. IEEE Transactions on Circuits and Systems II: Express Briefs. 2012;**59**: 491-495

Chapter 8

How Much Is the Cost of Implementing Arithmetic on a Quantum Computer?

Filippo Ghiglieno, Paulo Henrique Dias Ferreira, Vinicius Tribuzi and Olavo Leopoldino da Silva Filho

Abstract

This book chapter explores the transition from classical to quantum computing, emphasizing the capabilities and challenges associated with quantum bits (qubits). Unlike classical computing, where the information is represented as binary digits, known as bits, quantum bits, or qubits, although they share similarities with classical bits, such as 0 and 1 states, they operate in superposition, simultaneously encompassing both 0 and 1 states. This unique property allows quantum computers to perform certain computations, traditionally done sequentially on classical computers, more efficiently, in a single operation with a qubit. Through comparative analysis, we investigate the feasibility and costs of integrating quantum computing into daily applications, examining the potential for personal quantum computing devices. Our findings highlight significant advancements in computational speed of routine mathematical operations, although not yet economically viable or competitive in the current market. Future directions question the market readiness for quantum computing, suggesting a pivotal shift towards cloud-based quantum computing resources.

Keywords: classical logic and arithmetic, quantum logic and quantum arithmetic, electronic hardware, quantum hardware, scalability/adaptability

1. Introduction

"To Rachele and Mattia, the next users of this amazing technology."

Quantum mechanics came as a thunderstorm to the fundamental pillars in our classical understanding of the physical realm. Since its inception in the 1920s, it has enriched our comprehension significantly, marked by pioneering contributions from scientists like Niels Bohr, Werner Heisenberg, and Erwin Schrödinger, and has led to technological innovations that have become cornerstones of modern civilization. The theory has revolutionized the electronics and the information science with the transistor, the backbone of contemporary electronics [1, 2], and the advent of optical communication through lasers [3] are among the myriad contributions of quantum mechanics to technology and information science. Furthermore, recent advancements in quantum chemistry have facilitated the creation of novel materials [4] and pharmaceuticals [5], underscoring the theory's broad impact. Today, quantum technologies contribute significantly to the Gross Domestic Product (GDP) of numerous countries, embodying a substantial sector within the global economy [6]. Nonetheless, similar to the slow assimilation of Maxwell's equations in the nineteenth century, which required a century to fully grasp and catalyze technological advancements, it is to expect that the recent developments in experimental techniques for generating and controlling quantum states will lay the foundations for a potential second quantum revolution. This resurgence holds the promise of reshaping societal and economic landscapes once more. Quantum mechanics, initially a purely theoretical framework, underwent a transformative journey spanning over half a century. It transitioned from a mere physical theory to a cornerstone of our understanding and manipulation of information—the very foundation of modern information technology. The old Shannon's paradigm according to which the purely mathematical bit can be implemented in the physical word has now been replaced by Landauer's view, in which the information cannot be detached from its physical embodiment [7]. In 1982, the so-called "no-cloning theorem" confirmed new rules must be considered when information was quantically encoded [8]. In the same year, R. Feynman clearly predicted the quantum computer supremacy on the classical machines [9], which highlights the unique and transformative nature of quantum information processing. Subsequent milestones, including the first establishment of a quantum cryptography protocol published by Bennett and Brassard [10] in 1984, which allowed the creation of symmetric keys with perfect secrecy between the end points of a channel able to transmit quantum correlations, and the first, albeit primitive, implementation in 1989 [11] over a 25-cm free space, have added credibility to its claim of being a viable technology. This milestone effectively inaugurated the domain of quantum key distribution (QKD), marking a pivotal moment in its practical application, further exemplifying quantum mechanics' profound impact on communication security and computational theory. Nevertheless, the no-cloning principle, which gives the security of quantum key distribution (QKD) and quantum cryptography, also imposes restrictions on the faithful amplification of quantum signals. This limitation ultimately constrains the maximum achievable distance of QKD transmissions. It's worth noting that while the no-cloning principle prohibits the direct copying of quantum states, it does not rule out the potential for quantum repeaters. Quantum repeaters are devices capable of establishing quantum correlations over vast distances without resorting to state replication. Therefore, they offer a promising solution to overcome the inherent limitations of QKD, enabling it to transcend physical constraints such as distance and losses [12], overcoming its inherent limitations. The advent of quantum cryptography gained significant traction in 1994, notably propelled by Peter Shor's groundbreaking quantum algorithm [13]. This pivotal moment made computer scientists and cryptographers worldwide realize the potential threat posed by quantum computers to existing public key ciphers and cryptographic systems. It became evident that if quantum computers were successfully developed, they could potentially compromise the security of current cryptographic protocols. The main question became, after the development of quantum computers, how long it would take before the rise of cryptographic vulnerabilities. While a quantum computer boasting 53 qubits has, in mere minutes, tackled computations that traditional computers would need millennia to solve [14], quantum key distribution systems are already in a commercial stage and many companies and governments have invested heavily in researching quantum information technologies, which reflect the rapid progression and application potential of quantum technologies. Moreover, as we stand on the brink of a new technological era, the contributions of quantum mechanics to sensing, metrology, and beyond continue to expand the horizons of what is scientifically and practically achievable. Particularly, integrating quantum computers into daily applications notably involves leveraging quantum and reversible arithmetic techniques, advancing beyond traditional methods. This encompasses employing Draper's pioneering work on Quantum Fourier Transform (QFT)-based arithmetic [15], a critical foundation for quantum computation, alongside exploring modern advancements like ripple-array approaches [16]. These methodologies are essential for harnessing the unique computational capabilities of quantum systems, enabling more efficient processing and problem-solving strategies that significantly outperform classical computing paradigms in specific tasks. Despite existing challenges that must be addressed for its broader application, the impact of quantum technologies on our future technological landscape is undeniable. These advancements are poised to significantly influence various sectors, underscoring their emerging critical role in shaping the next technological era [17, 18]. While the potential for considerably faster computations is evident [19], the practical implications of integrating quantum hardware into everyday computing pose critical questions regarding cost-effectiveness and energy efficiency. This requires a shift in inquiry towards a more targeted analysis: "What is the comparative cost analysis between implementing quantum arithmetic operations versus classical operations in terms of computational resources and energy consumption?" This refined question underscores the need to evaluate the practical implications of quantum computing, focusing on the balance between its revolutionary computational speed and the logistical and economic challenges of adopting quantum hardware for everyday computing.

2. Classical and quantum arithmetic

2.1 Adding binary numbers

Binary addition operates on a straightforward principle, mirroring the conventional arithmetic process but within the binary numeral system. This approach ensures that adding two binary numbers adheres to specific rules, yielding results comparable to those of traditional arithmetic, albeit expressed in binary form. This foundational concept underscores the logical consistency between binary and standard arithmetic operations. By doing binary addition, two numbers in their binary representation are added to get the resultant number. This operation is typically performed using an adder, a fundamental component within the arithmetic logic unit. In the arithmetic logic unit, the adder serves as the foundational element for various arithmetic functions, including subtraction, multiplication, and division [20]. Algorithms for these operations are intricately linked to the principles of binary addition. The rules dictating binary addition are as follows.

2.1.1 The rules

We want to add two four bit numbers $A_3A_2A_1A_0$ and $B_3B_2B_1B_0$. We have to work through each column from the right to the left. In the following example, the first number is 1001 ($A_3 = 1$, $A_2 = 0$, $A_1 = 0$, $A_0 = 1$) and the second one is 1111 ($B_3 = 1$,

$B_2 = 1, B_1 = 1, B_0 = 1$). Starting with the rightmost bit, we first have to calculate 1 (A_0) plus 1 (B_0), and 1 (C_1) carries out to the next column.

Step 1 (S1):

$$\begin{array}{r} \\ 1001 \\ +\ 1111 \\ \hline 0 \end{array} \qquad \begin{array}{r} 1 \\ 1001 \\ +\ 1111 \\ \hline 0 \end{array}$$

Now, the second step consists in calculating the second column: 0 (A_1) plus 1 (B_1) and adding finally the carry 1 (C_1). Again 1 carries out to the next column (C_2). We advance further, in the third column we have 0 (A_2) plus 1 (B_2) and the carry 1 (C_2). Again 0 is the result with the carry 1 (C_3).

Step 2 (S2):

$$\begin{array}{r} 11 \\ 1001 \\ +\ 1111 \\ \hline 00 \end{array} \qquad \begin{array}{r} 111 \\ 1001 \\ +\ 1111 \\ \hline 000 \end{array}$$

In the last step, we have to sum 1 (A_3) and 1 (B_3) adding the carry 1 (C_3). The result is 1000 with one more carry 1 (C_4) for the next column.

Step 3 (S3):

$$\begin{array}{r} 1111 \\ 1001 \\ +\ 1111 \\ \hline 1000 \end{array} \qquad \begin{array}{r} 1111 \\ 1001 \\ +\ 1111 \\ \hline 1 \leftarrow 1000 \end{array}$$

Formally, we can summarize the sum protocol in the following way:

$$\begin{array}{r} C_4C_3C_2C_1 \\ A_3A_2A_1A_0 \\ +\quad B_3B_2B_1B_0 \\ \hline C_4 \leftarrow S_3S_2S_1S_0 \end{array}$$

2.1.2 The half adder: the circuit

We begin building a logical circuit that adds two bits following the added rules. We have four situations to face:

$$\begin{array}{lr} & 0 \\ + & 0 \\ \hline \textit{Sum} & 0 \\ \textit{Carry} & 0 \end{array} \qquad \begin{array}{lr} & 1 \\ + & 0 \\ \hline \textit{Sum} & 1 \\ \textit{Carry} & 0 \end{array}$$

	0		1
+	1	+	1
Sum	1	*Sum*	0
Carry	0	*Carry*	1

We name the input bits A and B, the sum S, and the carry bit C. The four situations can be summarized in a so-called truth table (see **Table 1**). The sum can be implemented with a logical XOR. The carry can be expressed with a logical AND.

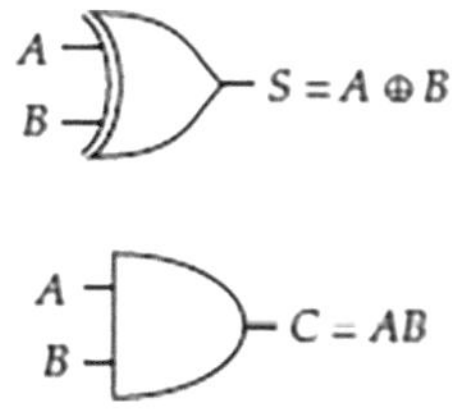

Figure 1 shows how the circuit, so-called *half adder*, works in practice.

2.1.3 The full adder: the circuit

For the next bits, the logical circuit is more complex, it is not enough for a half adder. We have to add them along with whatever was carried in. Since this allows for a carry, it is called *full adder*. We call the two input bits A and B. The bit we carry is C_{in}. We build again the truth table for understanding the gates we need to implement the logical operation:

From the truth table (**Table 2**), we can see that the carry out is 1 when A and B are 1, or when $C_{in} = 1$ and $A \oplus B = 1$:

$$C_{out} = AB + C_{in}(A \oplus B) \tag{1}$$

From the truth table, we also notice that the sum S is 1, whenever one of the inputs (**Figure 2**) A, B, or C_{in} is 1, or when all three are 1. This is equivalent to XOR of all the three bits:

$$S = A \oplus B \oplus C_{in} \tag{2}$$

It is now evident that to sum two four bit numbers, the circuit must have one half adder and three full adder. **Figure 3** shows how the circuit works in practice.

A	B	S	C
0	0	0	0
1	0	1	0
0	1	1	0
1	1	0	1

Table 1.
Truth table for the logical half adder.

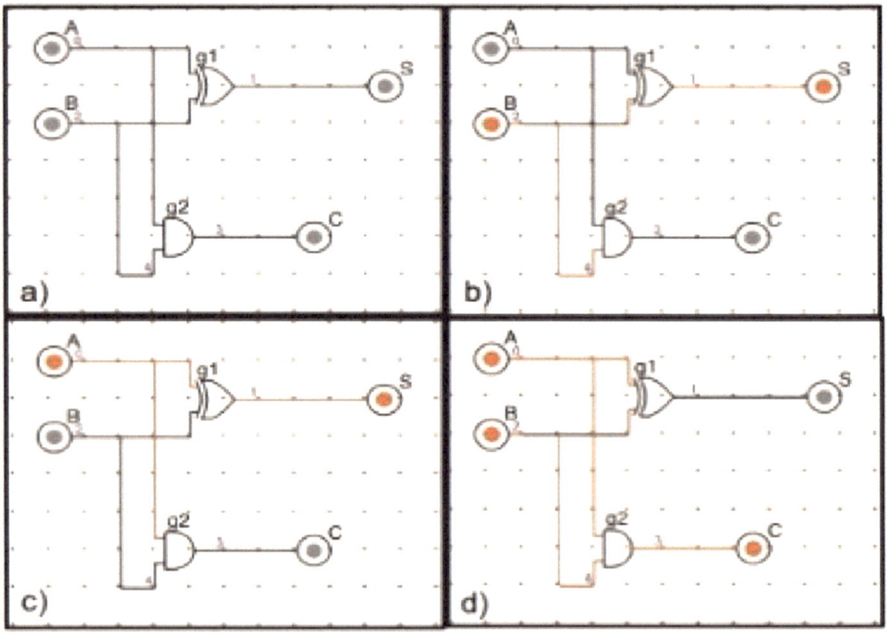

Figure 1.
Logical circuit for a half adder (electronic circuit simulation by Brian studio): (a) A = 0, B = 0; (b) A = 0, B = 1; (c) A = 1, B = 0; (d) A = 1, B = 1.

A	B	C_{in}	S	C_{out}
0	0	0	0	0
0	0	1	1	0
1	0	0	1	0
1	0	1	0	1
0	1	0	1	0
0	1	1	0	1
1	1	0	0	1
1	1	1	1	1

Table 2.
Truth table for the logical full adder.

3. Reversible and irreversible processes in physics

In the Resnick, Halliday, and Krane book [21], we can read this definition:

> *... In a reversible process, we make a small change in a system and its environment; by reversing that change, the system and its environment will return to their original conditions.*

The definitions in other introductory books are not much different or clearer [22, 23]. In thermodynamics, the principle of reversibility encompasses the concept

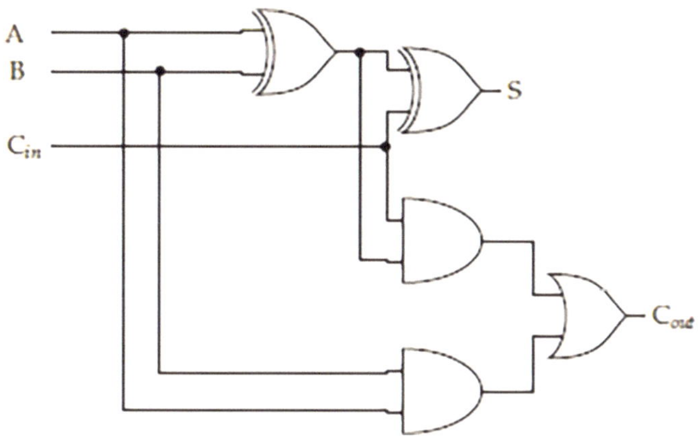

Figure 2.
Full adder circuit diagram.

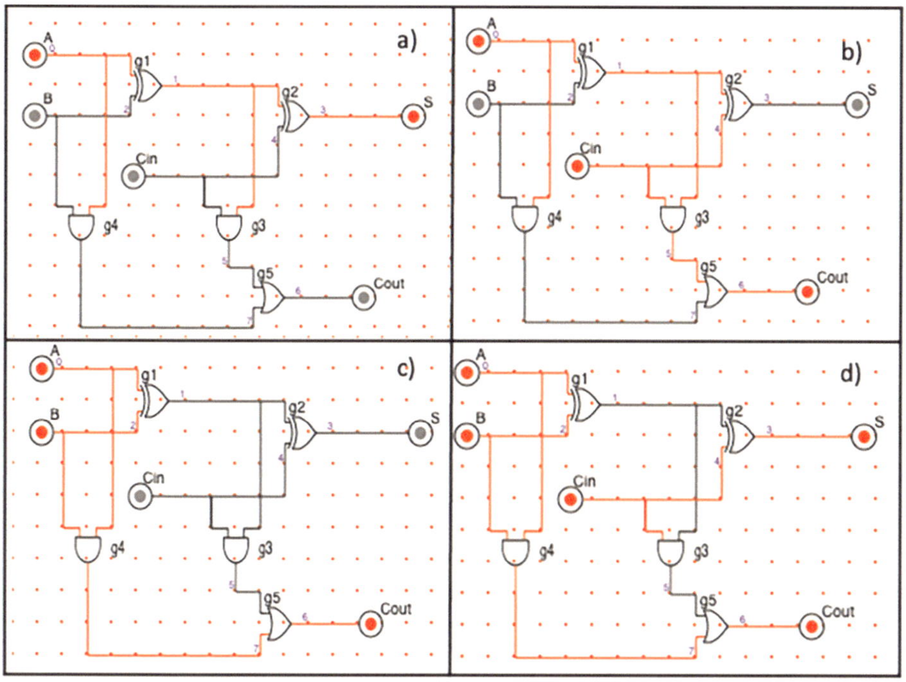

Figure 3.
Logical circuit for a full adder (electronic circuit simulation by Brian studio): (a) A = 0, B = 0; (b) A = 0, B = 1; (c) A = 1, B = 0; (d) A = 1, B = 1.

that certain processes are capable of being inverted, ensuring that a system can revert to its original state without exerting any lasting influence on other systems that have interacted with it (an ideal example of this process would be the singular swing of a frictionless pendulum). Irreversibility emerges because of the intricate interactions between a given system and other concurrent processes. Irreversible processes inherently involve dissipative phenomena, such as heat transfer across a finite temperature gradient or the manifestation of irreversible chemical reactions. In essence, the dichotomy between reversible and irreversible processes encapsulates the intricate dance of thermodynamic principles within the intricate choreography of the physical world. On the contrary, quantum mechanics and quantum circuits inherently embody reversibility, a consequence of the unitarity principle intrinsic to quantum mechanics. Based on the strict reversibility of the laws of microphysics, Landauer [24], Bennett [25], Priese [26], Fredkin and Toffoli [27], Feynman [28], and others envisioned a reversible computer that cannot allow any ambiguity in backward steps of a calculation. In this perspective, operations described by unitary operators, preserving purity and information, define the reversible transformations in quantum circuits.

3.1 Reversible gate

A reversible gate is a type of logic gate that allows the original input to be derived from its output. For instance, the NOT gate is an example of a reversible gate. Based on the truth table, the outputs of a NOT gate are distinct,

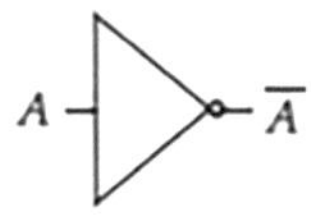

enabling the operation to be reversed. Hence, if the output from the gate is 1, the input must have been 0, and the reverse is also true. The reversibility of the NOT gate stems from the fact that the input can be always unequivocally determined (**Table 3**).

3.2 Irreversible gate

An irreversible gate is the opposite of a reversible gate. Given the output of the gate, we cannot determine unequivocally what the input is. An example is the AND gate (**Table 4**).

From the truth table, the output of the AND gate being 1 clearly indicates that both inputs are 1. However, if the output is 0, it does not uniquely specify the inputs, which

A	$\overline{A}$
0	1
1	0

Table 3.
Truth table for the NOT gate.

A	B	AB
0	0	0
1	0	0
0	1	0
1	1	1

Table 4.
Truth table for the AND gate.

A_{in}	B_{out}	C_{out}
0	0	1
1	1	0

Table 5.
Truth table for an irreversible gate.

A	B	C	D
0	0	1	1
1	0	1	0
0	1	0	1
1	1	0	0

Table 6.
Truth table for reversible gate.

could be 00, 01, or 10. Thus, from the output alone, the inputs to the AND gate cannot typically be determined, making it an irreversible gate. Note that the AND gate accepts two input bits, allowing for four possible combinations (00, 10, 01, and 11), but produces only one output bit, which can be either 0 or 1. The reduction in possible output states compared to input states means some information is lost, making it impossible to reliably deduce the inputs from the output. Moreover, in situations where there are more input bits than output bits, the circuit remains irreversible because some output scenarios do not correspond uniquely to specific inputs. Consider, for example, the following truth table (**Table 5**).

If both outputs are 0, the input remains unspecified according to the truth table; hence, its inverse is not fully determined. Therefore, a logic gate must have at least as many input bits as output bits to be reversible [29]. This is a necessary condition but not always sufficient. For instance, the gate described below is reversible (**Table 6**).

The gate is reversible because, given the output, it is always possible to determine the input. No information is lost since we can always recover the inputs from the outputs. Moreover, we have just seen, the AND gate is not reversible.

3.2.1 Making reversible gates irreversible

We consider a gate with inputs A and B and outputs $f(A, B)$, which is a function that output 0 or 1 depends on the inputs A and B. For example, for AND gate, the function would map $f(0,0) = 0, f(0,1) = 0, f(1,0) = 0$, and $f(1,1) = 1$. We can make it reversible with XORing output with a third input C (see **Table 7**).

In the rightmost column of the truth table, we used $f(A,B)\oplus 0 = f(A,B)$ and $f(A,B)\oplus 1 = \overline{f}(A,B)$. So this implements the original gate when $C = 0$. But the overall circuit is reversible. In the output, every permutation of A and B shows up twice, once with $f(A, B)$ and once with $\overline{f}(A,B)$, ensuring that the outputs are unique [30]. The technique can be generalized several ways. **Figure 4** shows how to implement the logical circuit for a generic two input ports/one output logical gate.

A	B	C	A	B	$f(A,B)\oplus C$
0	0	0	0	0	$f(0,0) \Rightarrow 0$
0	0	1	0	0	$\bar{f}(0,0) \Rightarrow 1$
0	1	0	0	1	$f(0,1) \Rightarrow 0$
0	1	1	0	1	$\bar{f}(0,1) \Rightarrow 1$
1	0	0	1	0	$f(1,0) \Rightarrow 0$
1	0	1	1	0	$\bar{f}(1,0) \Rightarrow 1$
1	1	0	1	1	$f(1,1) \Rightarrow 1$
1	1	1	1	1	$\bar{f}(1,1) \Rightarrow 0$

Table 7.
Truth table for the XORing irreversible gate.

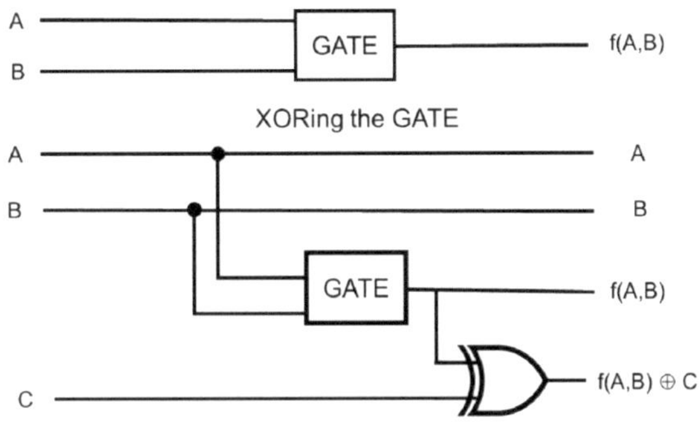

Figure 4.
XORing the GATE.

4. Quantum adder

4.1 Qubit

A quantum bit, often referred to as a qubit, shares similarities with and differs from classical bits. Like its classical counterpart, a qubit can assume either of the two values, 0 or 1. In the language of quantum physics, using Dirac notation, these states are represented by the kets:

$$|0\rangle = \begin{bmatrix} 1 \\ 0 \end{bmatrix} \qquad |1\rangle = \begin{bmatrix} 0 \\ 1 \end{bmatrix} \tag{3}$$

A qubit possesses the ability to store both classical and quantum information concurrently, a phenomenon called quantum superposition. Analogous to classical bits, where the number of states increases exponentially with the number of bits (e.g., 8 bits allowing representation of integers from 0 to $2^8 - 1$), qubits follow a similar pattern. Still, with 8 qubits, it is possible to simultaneously encode all integers within the range of 0 to $2^8 - 1$. Consequently, the memory capacity of a quantum computer grows exponentially with the number of qubits. This exponential growth implies that the number of bits required to represent a set of qubits also increases exponentially with the set length. A qubit in superposition can be expressed as a linear combination of computational basis states, where an arbitrary qubit $|\psi\rangle$ is defined as $|\alpha|^2 + |\beta|^2 = 1$. The numbers α and β, known as the probability amplitudes, weigh the probability of measuring a basis state. r. Those are complex numbers with the restriction of $|\alpha|^2 + |\beta|^2 = 1$.

4.2 Quantum gate: CNOT and Toffoli gate

Quantum gates act on qubits like logic gates act on bits. The controlled-NOT gate or CNOT inverts the right qubit if the qubit is 1:

$$CNOT\ |00\rangle = |00\rangle \tag{4}$$
$$CNOT\ |01\rangle = |01\rangle \tag{5}$$
$$CNOT\ |10\rangle = |11\rangle \tag{6}$$
$$CNOT\ |11\rangle = |10\rangle \tag{7}$$

In this arrangement, the qubit on the left is known as the control qubit, and the qubit on the right is termed the target qubit. It is important to note that the control qubit remains unchanged by the CNOT operation, while the target qubit is transformed into the XOR of the input values:

$$CNOT\ |a\rangle\ |b\rangle = |a\rangle\ |a \oplus b\rangle \tag{8}$$

Thus, the CNOT gate is a quantum XOR gate. We can implement the sum using two control not, the second CNOT turns the previous in:

$$CNOT\ |c\rangle\ (CNOT\ |a\rangle\ |b\rangle) = |c\rangle\ |a\rangle\ |a \oplus b \oplus c\rangle \tag{9}$$

This quantum circuit can implement the quantum sum [30]. It is clear that due to the reversibility in quantum computing, we have to pay for the *quantization* by including one more qubit at the input stage (the *ancilla* qubit) in comparison to the bits in the classical circuit.

The Toffoli gate is a three-qubit gate in quantum computing, also called Controlled-Controlled-NOT gate.

$$CCNOT|000\rangle = |001\rangle \tag{10}$$
$$CCNOT\ |001\rangle = |001\rangle \tag{11}$$
$$CCNOT|010\rangle = |010\rangle \tag{12}$$
$$CCNOT|011\rangle = |011\rangle \tag{13}$$
$$CCNOT|100\rangle = |100\rangle \tag{14}$$
$$CCNOT|101\rangle = |101\rangle \tag{15}$$
$$CCNOT|110\rangle = |111\rangle \tag{16}$$
$$CCNOT|111\rangle = |110\rangle \tag{17}$$
$$CCNOT|c\rangle\ |b\rangle\ |a\rangle = |a\rangle\ |b\rangle\ |ab \oplus c\rangle \tag{18}$$

Since the Toffoli gate is a quantum universal gate, quantum computers can efficiently convert classical algorithm utilizing Toffoli gates. This allows to create an AND of A and B, XORed with C:

$$CCNOT\,|c\rangle\,|a\rangle\,|b\rangle = |a\rangle\,|b\rangle\,|ab\oplus c\rangle \tag{19}$$

$$CNOT\,|a\rangle\,|b\rangle = |a\rangle\,|a\oplus b\rangle \tag{20}$$

4.3 Quantum adder implemented on the IBM platform

Quantum computing is emerging from the research in the laboratory onto the marketplace. Numerous enterprises are in the process of crafting prototype quantum processors. These early stage quantum processors, termed noise-intermediate-scale quantum (NISQ) devices, are characterized by their susceptibility to decoherence, making them not fault-tolerant devices, and they are of intermediate-scale nature since they feature a moderate quantity of qubits. Several companies have opened access to their preliminary quantum processors for experimental purposes. Pioneering this accessibility, IBM was the foremost to provide cloud-based access to their quantum processors via the internet. IBM's quantum computing resources, including their smaller quantum processors, are accessible to the public via their platform (https://quantum-computing.ibm.com), while larger and more advanced processors are offered through commercial access. This arrangement facilitates a range of engagement with quantum computing, from educational to research-oriented applications. The Circuit Composer on the website provides a drag-and-drop interface for the programming quantum circuits. Using the Circuit Composer, it is possible to create the circuits presented in the previous subsection, as shown in **Figure 5**. We note that q[0], q[1], and q[2] are the three qubit to

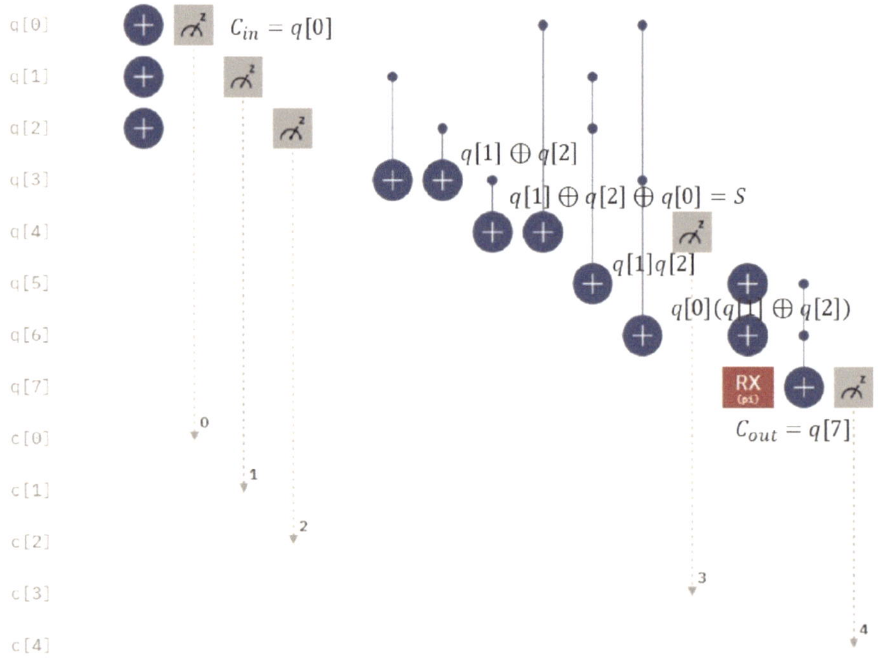

Figure 5.
Quantum adder on the IBM platform.

sum, q[0] is the carry in. The three not gates ⊕ initialize to 1 the input values. q[3] and q[4] are the *ancilla* qubits for making reversible, respectively, q[1] ⊕ q[2] and q[0] ⊕ q[1] ⊕ q[2]. The two Toffoli gates implement the logical AND between q[1], q[2] (q[1]q[2]) and between q[0], q[1] ⊕ q[2] (q[0][q[1] ⊕ q[2]]) with the support of the two *ancilla* qubits q[5] and q[6]. The last qubit q[7] produces C_{out} through an anti-Toffoli gate, implementing an OR gate between q[1]q[2] and $q[0](q[1]\oplus q[2]) = C_{out}$.

5. Results and discussion

Using the Circuit Composer, shown in **Figure 5**, it is possible to create the circuits presented in the previous subsection. Where q[0], q[1], and q[2] are the three qubits to sum, q[0] is the carry in. The three NOT gates initialize the input values to 1. On January 4th, 2024, we ran the quantum circuit 1024 times on IBM's IBM_Brisbane 13 processor, with 127 qubits available. The following table (see **Table 8**) summarizes

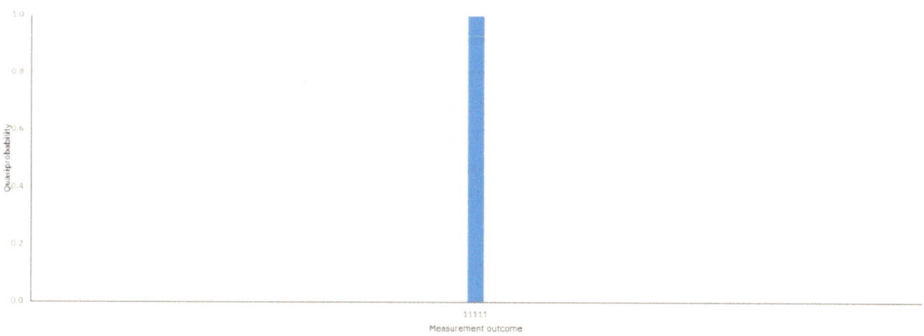

Figure 6.
Quantum adder quasi-probability histogram on the IBM QSAM (classical quantum) platform when q[0] = 1, q[1] = 1, q[2] = 1.

q[7]	q[5]	q[2]	q[1]	q[0]	Frequency
1	0	0	1	1	0.005963903511760564
1	0	1	0	1	0.005789115501746171
1	0	1	1	0	0.007055794279213567
1	0	1	1	1	0.13005895397294628
1	1	1	0	1	−0.003797515073072619
1	1	1	1	0	−0.01562920614909066
1	1	1	1	1	0.6528051438929247
0	0	0	1	1	0.0004680269228594772
0	0	1	1	0	−0.0007243258136245517
0	0	1	1	1	0.037259824909123135
0	1	1	1	0	−0.0024218131094713726
0	1	1	1	1	0.18317209715468513

Table 8.
Quasi-probability for the quantum adder implemented on the IBM platform.

the quasi-probability distribution for 12 outputs after 1024 trials in 17 seconds (in queue: 2 h 26 m 30.9 s) when q[0] = 1, q[1] = 1, q[2] = 1. Quasi-probability distributions are distinct from standard probability distributions as they permit negative values, yet their total must still equate to 1, akin to regular probabilities. This characteristic often emerges in the context of error mitigation strategies, illustrating the unique aspects of quantum computing calculations. Negative quasi-probabilities may appear when using error mitigation techniques [31]. The implemented protocol [32] excels in capturing correlated noise efficiently. It is designed to scale effectively for deployment in large quantum devices and serves as a demonstration of the concept of probabilistic error cancelation (PEC) on a superconducting quantum processor. PEC, as showcased in this implementation, is a technique adept at reversing well-characterized noise channels, leading to accurate and unbiased estimates of physical observables. The table shows the result q[4] = S = 1 and q[7] $= C_{out} = 1$ with 65% confidence, while q[4] = S = 1 and q[7] $= C_{out} = 0$ shows a quasi-probability next to the 18%. The IBM platform offers the possibility to run quantum circuit on an advanced quantum cloud-based computer simulator. We ran the same circuit on the ibmq_qsam_simulator, performing the simulation on a classic computer that resides on the cloud and simulates a 32-qubit computer. **Figure 6** indicates the result q[4] = S = 1 and q[7] $= C_{out} = 1$ with 100% confidence. The quantum adder circuit exhibits consistent results between classical and quantum hardware; however, the intricacy of the circuit renders qubit q[7] $= C_{out}$ susceptible to noise, with a frequency of 18%. In a subsequent experiment on the same day, the circuit was rerun with altered input values (q[0] = 1, q[1] = 0, q[2] = 1) on the identical platform and quantum hardware. Following 1024 trials and 18 seconds of utilizing the ibm_brisbane device, the outcomes were as follows: q[4] = S = 0 and q[7] $= C_{out} = 1$ with a quasi-probability of 0.426898811824693, and q[4] = S = 1 and q[7] $= C_{out} = 1$ with a quasi-probability of 0.341177111084729. This last result underlines the importance of noise reduction for the next generation of quantum computers and the need for the introduction of error-correction protocols in famous complex circuits like those employed in Shor's algorithm or Grover's algorithm [33]. Notably, the simulation using the ibmq_qsam_simulator classical cloud computer yielded q[4] = S = 0 and q[7] $= C_{out} = 1$ with a quasi-probability of 100% (see **Figure 6**). We executed a for loop 1024 times to implement the same logical classical operation consistently on a laptop with Intel(R) Core(TM) i7-8650U CPU@ 1.90 GHz-2.11 GHz and 16.0 GB RAM. The program in the MATLAB environment (*@Matworks*) when $A_1 = 1$ $A_0 = 1$ and $C_{in} = 0$ outputs the correct result with probability 100% in less than 1 millisecond (in average). In present-day classical computers, the probability of encountering errors is exceedingly low, often reported at the scale of $p \approx 10^{-18}$. In contrast, for cutting-edge quantum computers, the error probability is notably higher, typically ranging around 10^{-2} or, under optimal conditions, 10^{-3} [34]. The process of addressing errors in classical bit addition revolves around encoding redundant information from one bit across multiple bits. This ensures that if an error occurs in one of the redundant bits, the encoded data remain intact. A basic illustration of this concept involves a three-bit code employing a majority voting mechanism. To retrieve the encoded data, all three bits are measured, and the result is determined by the majority vote among them. Encoded information is determined through a majority vote among three bits. For instance, if a distortion changes the third bit in a sequence, transforming 111 to 110, the predominance of 1 s in the sequence 110 suggests the original encoded information was a 1. This method underscores the resilience of error-correction codes in maintaining data integrity despite errors (**Table 8**) [35].

5.1 Error correction

Correcting errors on quantum bits is indeed a more intricate challenge. This complexity posed a significant challenge for researchers in the early stages of quantum computing. The nuanced nature of quantum information and the susceptibility of qubits to environmental factors make error correction a substantial and ongoing focus of research in the field, especially for large and complex circuits. Overcoming these challenges is crucial for the development of reliable and scalable quantum computing technologies. The major obstacles thought to prevent quantum error are:

1. the no-cloning theorem [8];

2. measuring a quantum state causes it to collapse into an eigenstate of the measured observable [36];

3. within classical computing, a singular error type exists, the bit flipping, which toggles the bit between 0 and 1. In contrast, a quantum bit faces a multitude of potential errors, primarily due to various phase shifts, highlighting the complexity of quantum error correction [37].

In a bit flip channel, with three qubits it is possible to encode and protect the information, but it increases the hardware resources necessary for implementing the quantum circuit. To protect the information from the flip errors in a bit flip channel (the X gate), it is possible to encode the qubit using three qubits, not cloning, but applying the CNOT gate:

$$|\psi\rangle = \alpha|0\rangle + \beta|1\rangle \rightarrow \alpha|000\rangle + \beta|111\rangle = |\psi_3\rangle \tag{21}$$

If we perform the parity measurement both on qubits 1,2 and 2,3, i.e., measure the product $\sigma_{z_1} \otimes \sigma_{z_2}$ and the product $\sigma_{z_1} \otimes \sigma_{z_2}$ (where σ_{z_i} is the Z 2 x 2 Pauli matrix acting on the qubit i):

$$\sigma_{z_1} \otimes \sigma_{z_2} \ |\psi_3\rangle, \tag{22}$$

$$\sigma_{z_2} \otimes \sigma_{z_3} \ |\psi_3\rangle, \tag{23}$$

the carried information is not changed, the quantum state $|\psi_3\rangle$ just gains a global phase factor marking a possible flip. In this case, applying an X gate to this qubit (σ_{x_3} 2 x 2 Pauli matrix), it flips back to the original state. The protocol works also when the rotation is not a flip (X gate), but an arbitrary rotation ϑ (*RX* gate on the IBM platform). In quantum calculations (or computing), the role of stabilizers in error correction, as above, can be elucidated by examining how a quantum state $|\psi\rangle$ interacts with a unitary operator U. Specifically, if the application of U to $|\psi\rangle$ results in no change to the state ($U|\psi\rangle = |\psi\rangle$), then $|\psi\rangle$ is considered to be stabilized by U, indicating a foundational aspect of understanding error-correction codes through stabilizers. Implementing then is crucial in minimizing the likelihood of a single-qubit error propagating into a genuine error within an encoded single-logical-qubit state, which is composed of multiple physical qubits. However, this initial step towards achieving error resilience in quantum computing comes at a cost. The incorporation of error correction inevitably escalates circuit complexity and adds to the computational overhead. Moving beyond this challenge, the realization of fault-tolerant quantum computing necessitates the development of robust schemes and allocation of resources for executing logical operations and measurements on logical qubits in a manner that is inherently resilient to errors. While delving into the intricacies of achieving fault tolerance is beyond the scope of this chapter, it is indeed possible and represents a pivotal aspect of advancing quantum computing capabilities.

6. Conclusions

Quantum computation stands as a groundbreaking paradigm with the potential to significantly reduce complexity in mathematical problem-solving. Contemporary accessibility to quantum hardware through cloud architectures, as exemplified by platforms such as Quantum IBM on IBM and AWS on Amazon, presents varied pricing plans in terms of U.S. dollars per second. However, adhering to fundamental principles in quantum mechanics, quantum circuits inherently necessitate reversibility, leading to an augmented demand for basic resources (qubits). In this investigation, we explored the hardware cost implications of implementing a classical logical circuit, specifically the full adder, on an IBM quantum computer. In contrast to its classical counterpart, the quantum circuit demands six input qubits (three input values q[0],q[1],q[2] and three ancillas q[3],q[5],q[6]) and two output qubits (q[4] = S and q[7] = C_{out}), despite both circuits addressing a binary logic operation with three input bits (A_0 A_1 and C_{in}) and two output bits (S and C_{out}). The classical logic circuit can be effortlessly executed on commonplace calculators or personal laptops, offering straightforward implementation, reliable results, and swift outputs within milliseconds for a thousand trials. While the classical logic circuit seamlessly executes on standard calculators or personal laptops, featuring a straightforward implementation, dependable results, and rapid outputs within milliseconds for a thousand trials, the quantum counterpart introduces additional complexity. The quantum circuit, despite its higher hardware demands, is subject to the distinctive principles of quantum mechanics, and its execution involves the utilization of ancillary qubits for specific functionalities. This disparity highlights the inherent distinctions between classical and quantum computing paradigms, underscoring the need for a nuanced consideration of resource requirements when transitioning from classical to quantum implementations of logical circuits. Nowadays, the most common and efficient business solution on the market is shared quantum hardware in cloud architecture, but subject to queuing delays. Results are obtained within seconds for a thousand trials, yet they are influenced by noise and temporal stability issues. In this perspective, for routine day-to-day mathematical operations, quantum calculators or laptops do not presently emerge as economically viable, reliable, scalable, user-friendly, or competitive products in the market. Future prospects may witness alternative platforms based on diverse quantum technology approaches, such as photonic or ion trap-based circuits, presenting compelling business models for personalized quantum products. At present, the considerable expense associated with superconducting quantum bits (at approximately $10,000 [38]) necessitates the introduction of quantum facilities into the market primarily in the form of shared resources accessible via the internet, adopting time-of-use plans priced in U.S. dollars per second.

Acknowledgements

The authors would like to gratefully acknowledge the financial support from FAPESP, Brazil (São Paulo Research Foundation – Processes number: CEPID 2013/00793-6, 2018/11283-7 and AUXPE 2013/21569-1), CAPES (Coordenação de Aperfeiçoamento de Pessoal de Nível Superior), CNPq (Conselho Nacional de Desenvolvimento Científico e Tecnológico) Brazil, Universidade Federal de São Carlos (UFSCar), and Instituto de Física (IF) at the University of Brasilia (UnB).

Conflict of interest

The authors declare no conflict of interest.

Author details

Filippo Ghiglieno[1,2*†], Paulo Henrique Dias Ferreira[1†], Vinicius Tribuzi[1†] and Olavo Leopoldino da Silva Filho[2†]

1 Department of Physics, Federal University of São Carlos, São Carlos, SP, Brazil

2 Physics Institute, University of Brasilia, Brasilia (DF), Brazil

*Address all correspondence to: filippo.ghiglieno@df.ufscar.br

† These authors contributed equally.

References

[1] Bardeen J, Brattain WH. The transistor, a semi-conductor triode. Physics Review. 1947;**74**:230-231

[2] Bardeen J. Surface states and rectification at a metal semiconductor contact. Physics Review. 1948;**71**: 383-388

[3] Maiman TH. Stimulated optical radiation in ruby. Nature. 1960; **187**(4736):494

[4] Kl P, Barkoutsos F, Gkritsis J, Ollitrault PJ, Sokolov IO, Woerner S, et al. Quantum algorithm for alchemical optimization in material design. Chemical Science. 2021;**12**:4345-4352

[5] Tautermann CS. Current and future challenges in modern drug discovery: Quantum mechanics in drug discovery. Method in Molecular Biology. 2020;**2114**: 1-17

[6] Deloitte Insights. Available from: https://www2.deloitte.com/us/en/insights/topics/innovation/-computing-business-applications.html [Accessed: Jan 01, 2024]

[7] Landauer R. Information is physical. Physics Today. 1991;**44**(5):23-29

[8] Wootters W, Zurek W. A single quantum cannot be cloned. Nature. 1982; **299**:802-803

[9] Feynman RP. Simulating physics with computers. International Journal of Theoretical Physics. 1982;**21**(6):467-488

[10] Brassard G, Bennett CH. Quantum cryptography: Public key distribution and coin tossing. Theoretical Computer Science. 2014;**560**(1):7-11

[11] Bennett CH, Brassard G. The dawn of a new era for quantum cryptography: The experimental prototype is working! SIGACT News. 1989;**20**(4):78-80

[12] Briegel HJ, Dür W, Cirac JI, Zoller P. Quantum repeaters: The role of imperfect local operations in quantum communication. Physical Review Letters. 1998;**81**(26):5932-5935

[13] Shor PW. Algorithms for quantum computation: Discrete logarithms and factoring. In: Proceedings 35th Annual Symposium on Foundations of Computer Science, Santa Fe, NM, USA. 1994. pp. 124-134

[14] Arute F, Arya K, Babbush R, et al. Quantum supremacy using a programmable superconducting processor. Nature. 2019;**574**:505-510

[15] Draper TG. Addition on a quantum computer. arXiv/quant-ph/0008033. 2000

[16] Wang F, Luo M, Li H, Qu Z, Wang X. Improved quantum ripple-carry addition circuit. Science China Information Sciences. 2016;**59**: 042406

[17] Corcoles AD, Kandala A, Javadi-Abhari A, Douglas T. Challenges and opportunities of near-term quantum computing systems. Proceedings of the IEEE Science. 2020;**108**(8): 1338-1352

[18] Hassija V, Chamola V, Saxena V, Chanana V, Parashari P, Mumtaz S, et al. Present landscape of quantum computing. IET Quantum Communication. 2020;**1**(2):42-48

[19] Castelvecchi D. IBM releases first-ever 1,000-qubit quantum chip. Nature. 2023;**624**:238

[20] Kann CW III. Digital circuit projects – an overview of digital circuits through implementing integrated circuits. 1st ed. LibreTexts; 2023. Available from: https://eng.libretexts.org/Bookshelves/ [Accessed: 26 Jan 01, 2024]

[21] Resnick R, Hallyday D, Krane K. Physics. 5th ed. New York. Wiley; 2002

[22] Young HD, Freedman RA, Ford AL. Sears and Zemansky's University Physics. 11th ed. New York, USA: Addison-Wesley; 2004

[23] Giancoli D. Physics for Scientists and Engineers. 3th ed. Upper Saddle River, New Jersey: Prentice Hall; 2000

[24] Landauer RW. Irreversibility and heat generation in the computing process. IBM Journal of Research and Development. 1961;**5**(3):183-191

[25] Bennet CH. Logical reversibility of computation. IBM Journal of Research and Development. 1973;**17**:183-191

[26] Priese L. On a simple combinatorial structure sufficient for sublying nontrivial self-reproduction. Journal of Cybernetics. 1976;**5**:101-137

[27] Fredkin E, Toffoli T. Conservative logic. International Journal of Theoretical Physics. 1982;**21**:219-253

[28] Feynman R. Quantum mechanical computers. Optics News. 1985;**11**: 11-20

[29] Morita K. Theory of reversible computing. Monographs in Theoretical Computer Science. An EATCS series Springer Tokio Edition. 2017

[30] Wong T. Introduction to Classical and Quantum Computing. Rooted Grove Edition. 1st ed. 21 Jan 2022

[31] Berg LK, Minev Z, Kandala A, Temme K. Probabilistic error cancellation with sparse Pauli–Lindblad models on noisy quantum processors. Nature Physics. 1985;**19**:11-20

[32] IBM blog. With fault tolerance the ultimate goal, error mitigation is the path that gets quantum computing to usefulness. Available from: https://research.ibm.com/blog/gammabar-for-quantum-advantage [Accessed: Feb. 05, 2024]

[33] Grover LK. A fast quantum mechanical algorithm for database search. In: Proceedings of the Twenty Eighth Annual ACM Symposium on Theory of Computing Association for Computing Machinery, Philadelphia, Pennsylvania, USA. 1996. pp.212–219

[34] Kockum AF, Soro A, et al. Lecture notes on quantum computing. arXiv/quant-ph/2311.08445v1. 2024

[35] Glover N, Dudley T. Practical Error Correction Design for Engineers. 2th ed. CO, USA: Cirrus Logic; 2000

[36] Cohen-Tannoudji C, Diu B, Laloë F. Quantum Mechanics: Basic Concepts, Tools, and Applications. 2nd ed. Vol. 1. Hoboken, NJ, USA: Wiley-VCH; 2009

[37] Nielsen MA, Chuang IL. Quantum Computation and Quantum Information. 4th ed. Cambridge UK: Cambridge University Press; 2004

[38] Koestier J. Million-qubit quantum computing? How SEEQC plans to scale quantum computers. Available from: https://johnkoetsier.com/quantum-computing-seeqc/ [Accessed: Feb 05, 2024]

Chapter 9

Predictive Software Engineering: Delivering Effective Business Solutions through Custom Software Development

Boris Kontsevoi

Abstract

This paper explores the seven core principles of the Predictive Software Engineering (PSE) framework. These principles are designed to empower custom software development companies to deliver transparent and reliable solutions, all while adhering to predetermined budgets. The paper delves into each of the seven principles: Meaningful Customer Care, Transparent End-to-End Control, Proven Productivity, Efficient Distributed Teams, Disciplined Agile Delivery Process, Measurable Quality Management and Technical Debt Reduction, and Sound Human Development.

Keywords: agile, disciplined agile delivery, distributed team, predictive software engineering, measurable quality management and technical debt reduction system (MQM&TDR)

1. Introduction

Traditional software development often suffers from unpredictable elements and subjective decision-making. Predictive Software Engineering (PSE) addresses this issue by introducing a set of seven principles, informed by over 29 years of experience. These principles guide development teams towards a more standardized, controlled, and transparent approach, transforming software creation into a precise and predictable engineering endeavor.

2. Meaningful customer care

Customer-centricity remains the cornerstone of success for any business, regardless of its market orientation (B2B, B2C, etc.). Customer satisfaction, brand loyalty, and ultimately, customer acquisition volume collectively define an organization's success. When an organization's offerings fail to resonate with target audiences and translate into commercial success, it signifies that it has not yet maximized its potential.

Therefore, to objectively assess progress towards organizational goals, measuring customer satisfaction and feedback through key performance indicators (KPIs) becomes paramount. Should these metrics reveal an unfavorable trajectory, the focus should promptly shift towards transforming dissatisfied customers into ardent brand advocates. This objective can be achieved through the implementation of eight key strategies:

1. Streamline the onboarding process.
2. Offer a dedicated customer portal for easy access and engagement.
3. Establish a clear governance model for efficient issue resolution.
4. Define a transparent escalation path for handling complex concerns.
5. Implement early problem discovery methods to proactively address issues.
6. Conduct comprehensive analysis of customer complaints to identify root causes.
7. Proactively manage customer satisfaction through consistent monitoring and engagement.
8. Set realistic expectations to avoid disappointment and build trust.

3. Transparent end-to-end control

Transparency, characterized by openness, trust, and cooperation, remains a scarce commodity in the contemporary business landscape. Despite its relative rarity, achieving transparency offers significant advantages. Stakeholders, particularly clients, perceive transparent organizations as less risk-prone and more reliable entities, fostering a collaborative environment conducive to long-term success. This paper explores the multifaceted approach required to cultivate transparency within organizations, both internally and externally.

3.1 External transparency

- *Governance and project meetings*: Regularly convening governance and project status meetings with external stakeholders serves as a key mechanism for maintaining information dissemination and fostering trust. These forums provide a platform for open communication, allowing clients to actively engage with the organization and address any concerns promptly.

- *Agile methodologies*: The implementation of Agile methodologies plays a crucial role in promoting real-time visibility into project progress. This iterative approach enables stakeholders to remain informed and engaged throughout the development process, fostering adaptability and collaborative problem-solving.

- *Collaboration tools and project portals*: The effective utilization of technology-driven solutions such as teamwork collaboration tools and project portals empowers clients with real-time access to project details. This fosters a sense of ownership and transparency, enhancing stakeholder confidence in the organization's commitment to open communication.

- *Process audits and proactive monitoring*: Implementing regular process audits allows organizations to proactively identify and address potential roadblocks. Additionally, shifting from reactive monitoring to a proactive approach ensures issues are identified and addressed before they escalate, demonstrating a commitment to continuous improvement and mitigating potential risks for stakeholders.

- *On-site visits*: Scheduling on-site visits at key project milestones serves as a powerful tool for strengthening open communication with clients. These visits demonstrate the organization's commitment to transparency and collaboration, fostering trust and building stronger relationships with external stakeholders.

3.2 Internal transparency

- *Leadership engagement*: Fostering a culture of transparency necessitates active and consistent engagement from the organization's leadership. Conducting regular meetings with C-level executives facilitates clear communication and alignment across all organizational levels, ensuring all stakeholders are informed and aligned with company goals.

- *Departmental check-ins*: Establishing regular departmental check-in meetings with directors fosters information flow and collaboration across various divisions. This practice promotes transparency by breaking down silos and ensuring all departments are working towards common objectives.

- *Project management engagement*: Regular check-ins with project managers serve a two-fold purpose. Firstly, they ensure that project progress is closely monitored, and potential issues are addressed swiftly. Secondly, these interactions foster a transparent environment where project managers feel comfortable voicing concerns and seeking guidance from leadership.

- *Talent management integration*: Including the talent management team in fostering collaborative work environments underlines the organization's commitment to internal transparency. This collaborative approach empowers employees to voice concerns and actively participate in decision-making processes, promoting a sense of ownership and trust within the organization.

- *Team building and company events*: Investing in team-building events and company gatherings strengthens internal communication and interpersonal relationships within the organization. These events provide informal spaces for information exchange and collaboration, fostering a sense of trust and transparency among employees.

Cultivating a culture of transparency requires a comprehensive and multifaceted approach. By implementing the strategies outlined above, both internally and externally, organizations can establish themselves as reliable and trustworthy partners. This commitment to openness fosters trust with stakeholders, facilitates collaboration, and ultimately positions the organization for long-term success in today's competitive business environment.

4. Proven productivity

Evaluating software engineer productivity remains a persistent challenge across diverse organizations. This paper explores two prominent tools employed to gauge productivity and velocity during development:

- Measurable Quality Management and Technical Debt Reduction System (MQM&TDR) (addressed in a separate section)
- Software Development Efficiency Key Performance Indicator (KPI)

The Software Development Efficiency KPI encompasses ten sub-KPIs designed to assess various development aspects [1]:

- *Static code analysis (project & individual)*: Evaluates overall and individual developer code quality through automated tools.
- *Planned vs. actual work*: Compares planned workload with actual completion, providing insights into project planning efficiency.
- *Bugs per line of code (LOC)*: Measures defect frequency relative to code volume, indicating code quality and potential rework effort.
- *Opened bugs per feature*: Assesses defect prevalence within individual features, aiding in identifying areas for improvement.
- *Templates/algorithms/standard libraries missed*: Evaluates adherence to established practices and best practices utilization.
- *Missed exception/error handling*: Assesses implementation of robust error handling mechanisms, indicating code robustness and user experience.
- *Security problems*: Identifies potential security vulnerabilities within the code, highlighting critical areas requiring immediate attention.

To ensure effectiveness and integrity of the Software Development Efficiency KPI, several key guidelines are recommended:

- *Double code review*: Two developers independently review the same code for accuracy and comprehensiveness.
- *Documented code review*: Meticulously document the entire code review process for auditability and future reference.
- *Management review and approval*: Team managers or leaders should comprehensively review and approve the KPI results to ensure validity and alignment with organizational goals.

- *Anti-cheating tools*: Utilize tools that mitigate manipulation and simplify the review process.

- *Code review volume limitation*: Review smaller code segments (ideally less than 500 lines) to enhance focus and thoroughness during assessment.

5. Efficient distributed teams

The dissolution of geographical constraints has enabled organizations to embrace distributed teams, fostering the creation of a geographically diverse workforce. This model offers several advantages, including:

- *Time effectiveness*: Distributed teams, spanning various time zones, ensure continuous coverage. By strategically placing development centers, organizations can guarantee at least one team is always operational, particularly valuable for entities requiring 24/7 support.

- *Cost savings*: Hiring from diverse locations allows organizations to potentially reduce costs associated with salaries, rentals, and operating fees.

- *Talent pool diversification*: Distributed teams grant access to a virtually limitless pool of highly skilled talent, broadening the candidate pool beyond geographical limitations.

However, physical distance can introduce challenges such as communication issues and decreased efficiency. To ensure optimal performance within distributed teams, adhering to specific guidelines is crucial:

- *Task design*: Tasks should be meticulously designed to facilitate collaboration, considering factors such as time zone differences, industry regulations, and individual workloads.

- *Role clarity*: Roles and responsibilities of each team member should be clearly defined and communicated to avoid confusion and overlap.

- *Effective communication*: Utilizing established and well-defined communication strategies, such as closed-loop communication, is paramount.

- *Loose coupling*: Tasks should be loosely coupled, minimizing dependency on individual members' progress, and fostering overall team agility.

- *Mutual performance monitoring*: While maintaining individual responsibilities, team members can mutually monitor performance, fostering a sense of accountability and collaboration [2].

Beyond these considerations, social aspects also require attention. Team members from diverse backgrounds may exhibit differing communication styles and cultural norms. There are measures for addressing this:

- *Effective leadership*: Providing strong and effective leadership is critical. Leaders should be adept at motivating team members, fostering cohesion, and offering necessary coaching.

- *Shared experiences*: Cultivating common ground among team members is essential. When feasible, conducting shared training sessions in a centralized location can further enhance team unity.

By recognizing the benefits and adhering to these best practices, organizations can effectively leverage distributed teams to achieve their business objectives and harness the advantages of a geographically diverse workforce.

6. Disciplined agile delivery process

While Agile methodologies remain the predominant framework for software development, a vast array of adaptations exist. This paper posits that Disciplined Agile Delivery (DAD) emerges as a particularly advantageous variant for service organizations.

Unlike traditional Agile approaches, DAD employs a distinct lifecycle structured into three primary phases: inception, construction, and transition [3].

6.1 Inception phase

- *Vision definition*: The project vision is meticulously defined, establishing a clear roadmap for development efforts.

- *Stakeholder alignment*: Collaboration with stakeholders is crucial, ensuring project vision aligns with their expectations and objectives.

- *Initial planning*: This phase encompasses crafting the initial project plan, outlining the preliminary technical strategy, and identifying core requirements.

- *Team formation*: The initial project team is assembled, bringing together the necessary expertise and resources.

6.2 Construction phase

- *Adapting to change*: Continuous adaptation to evolving stakeholder needs remains a central focus during this phase.

- *Iterative development*: The focus shifts towards delivering a potentially usable solution through iterative development cycles.

- *Deployment readiness*: Efforts are directed towards achieving a deployable release state.

- *Quality assurance*: Maintaining or enhancing existing quality levels remains paramount throughout the construction phase.

- *Risk mitigation*: Addressing the most significant project risks becomes a key priority during this phase.

6.3 Transition phase

- *Production readiness*: This phase ensures the developed solution is fully prepared for deployment into the production environment.

- *Stakeholder preparation*: Equipping stakeholders with the necessary knowledge and resources to effectively receive the solution is crucial.

- *Deployment execution*: The final stage involves the successful deployment of the solution into production, marking the culmination of the project lifecycle.

By embracing the distinct phases and tailored approach of DAD, service organizations can leverage the agility and adaptability inherent in Agile methodologies while addressing the specific needs and challenges they encounter within their domain.

7. Measurable quality management and technical debt reduction

The Measurable Quality Management and Technical Debt Reduction System (MQM&TDR) emerges as a comprehensive framework designed to assess various facets of software development projects [4]. Its core functionalities encompass:

- *Technical debt measurement*: MQM&TDR facilitates the quantification of technical debt, enabling proactive management and mitigation strategies to prevent accumulation and potential future issues.

- *Software quality assessment*: The system provides an unbiased evaluation of software quality across various dimensions, encompassing source code quality, usability, security, performance, and more.

- *Business efficiency determination*: MQM&TDR aids in determining the product's business efficiency, offering valuable insights for stakeholders.

The benefits of MQM&TDR extend beyond technical teams, offering advantages to various stakeholders:

- *Testers and developers*: Gain objective feedback on their work, enabling continuous improvement and quality enhancement.

- *Managers*: Receive insights into the project's overall capabilities, facilitating informed decision-making.

- *Users*: Benefit from a reliable and well-performing final product.

- *Investors*: Utilize MQM&TDR's findings to assess investment risks and determine fair market value with greater accuracy.

The system's evaluation encompasses a comprehensive spectrum of software components:

1. Source Code Quality

2. Usability

3. Security

4. Performance

5. Business Logic

6. Solution Architecture

7. Data Quality

8. Open-Source Code Use

By leveraging MQM&TDR, various stakeholders can benefit from the following:

- *Proactive technical debt management*: MQM&TDR enables the identification and mitigation of technical debt before it becomes a significant burden.

- *Reduced costs*: Redevelopment and support costs can be significantly minimized through proactive quality management.

- *Informed progress evaluation*: MQM&TDR facilitates a data-driven approach to progress evaluation, fostering informed decision-making.

- *Predicted business efficiency*: The system provides insights into the product's anticipated business efficiency, allowing for strategic planning.

- *Comprehensive quality analysis*: MQM&TDR offers a holistic assessment of software quality, encompassing various critical aspects.

- *Detailed feature analysis*: The system provides in-depth analysis of key product features, aiding in refinement and optimization.

- *Compliance verification*: MQM&TDR helps ensure adherence to relevant standards and regulations.

- *Actionable improvement recommendations*: The system offers specific recommendations for improvement, guiding ongoing development efforts.

Given its multifaceted functionalities and demonstrably positive impacts, MQM&TDR presents a compelling approach for enhancing software development projects across diverse organizations.

8. Sound human development

While the language of human resources often frames individuals as mere resources, a more nuanced perspective recognizes employees as autonomous agents. These agents possess the capacity for independent decision-making, skill development, and goal setting. By prioritizing human development, organizations can cultivate not only a more skilled workforce but also a more fulfilled and loyal employee base [5].

This paper explores several key strategies for promoting human development within organizations:

a. *Benefits:* Expanding beyond standard benefits (e.g., medical insurance, compensation, office amenities) to incorporate training and certification programs alongside language learning opportunities (e.g., English courses) demonstrates a commitment to employee development.

b. *Mentorship programs:* Implementing internship and mentorship programs offers numerous benefits. Interns gain invaluable practical experience working on real-world projects, transcending purely theoretical knowledge. This exposure extends beyond technical skills in programming and testing, encompassing communication and process understanding. Moreover, successful mentorship programs can often lead to retention, with interns transitioning into permanent roles after the program concludes, fostering a sense of team continuity and internal knowledge transfer.

c. *Performance review and career growth:* Establishing individual development plans (IDPs) for each employee is crucial. Through collaborative sessions involving the employee, talent management personnel, and relevant team members, organizations can gain insights into employee aspirations and career goals. Leveraging these insights, career paths can be charted, outlining potential developmental opportunities and advancement possibilities. Furthermore, implementing regular performance reviews using a combined approach is recommended. This approach involves self-assessment by employees followed by managerial feedback, culminating in a collaborative discussion to identify areas for improvement and establish actionable steps for the following review cycle.

d. *Skills improvement:* Organizations must demonstrate a continuous commitment to facilitating skill development among their employees. This commitment can be manifested through three key strategies:

- *Engaging external experts:* Seeking external guidance from domain specialists allows employees to access cutting-edge knowledge and industry best practices.

- *Providing internal and online training:* Offering a diverse range of internal training programs and e-learning resources empowers employees to pursue self-directed learning and upskilling opportunities.

- *Establishing Centers of Excellence (CoEs):* Creating dedicated CoEs within the organization fosters knowledge sharing and expertise concentration, allowing employees to learn from industry leaders and internal specialists within specific domains.

By implementing these comprehensive strategies, organizations can move beyond the limitations of traditional human resource management and embrace a human-centric approach that prioritizes individual development, career advancement, and overall employee well-being. This shift in perspective fosters a more engaged, motivated, and skilled workforce, ultimately contributing to organizational success and sustainability.

9. Risks covered with PSE

By embracing Predictive Software Engineering (PSE) and adhering to the outlined principles, software development teams can effectively mitigate a range of potential risks, including:

- *Executive disengagement*: The principles of Transparent Control and Meaningful Care foster open communication and stakeholder buy-in, minimizing the risk of insufficient executive support.

- *Scope creep*: Transparent Control promotes clear and well-defined project scope, reducing the likelihood of ambiguity and uncontrolled expansion.

- *Inaccurate estimates*: Disciplined Agile methodologies encourage iterative development and continuous estimation refinement, mitigating the risk of inaccurate initial estimates.

- *Resource constraints*: Distributed Teams empower organizations to leverage a wider talent pool, potentially addressing resource limitations.

- *Skill gaps*: Sound Human Development practices, like training programs and skill-building initiatives, equip teams with the necessary skills and knowledge, minimizing the potential for inadequate training.

- *Unclear decision-making*: Meaningful Care emphasizes open communication and collaborative decision-making, reducing the risk of ambiguity and confusion arising from unclear decisions.

- *Infrastructure delays*: Transparent Control promotes proactive planning and communication, allowing for the timely identification and mitigation of potential infrastructure-related delays.

Through the adoption of PSE and its associated principles, software development teams can navigate these common challenges and increase the likelihood of project success.

10. Conclusion

Custom software development often encounters bottlenecks hindering efficient and reliable delivery. Predictive Software Engineering (PSE) emerges as a novel framework designed to address these challenges and reconstruct a more robust approach to software development service delivery.